The Biology of Soil

BIOLOGY OF HABITATS

Series editors: M. J. Crawley, C. Little,
T. R. E. Southwood, and S. Ulfstrand

The intention is to publish attractive texts giving an integrated overview of the design, physiology, ecology, and behaviour of the organisms in given habitats. Each book will provide information about the habitat and the types of organisms present, on practical aspects of working within the habitats and the sorts of studies which are possible, and will include a discussion of biodiversity and conservation needs. The series is intended for naturalists, students studying biological or environmental sciences, those beginning independent research, and biologists embarking on research in a new habitat.

The Biology of Soil

A Community and Ecosystem Approach

Richard D. Bardgett

Institute of Environmental and Natural Sciences,
Lancaster University

OXFORD
UNIVERSITY PRESS

OXFORD

UNIVERSITY PRESS

Great Clarendon Street, Oxford OX2 6DP

Oxford University Press is a department of the University of Oxford.
It furthers the University's objective of excellence in research, scholarship,
and education by publishing worldwide in

Oxford New York

Auckland Cape Town Dar es Salaam Hong Kong Karachi
Kuala Lumpur Madrid Melbourne Mexico City Nairobi
New Delhi Shanghai Taipei Toronto

With offices in

Argentina Austria Brazil Chile Czech Republic France Greece
Guatemala Hungary Italy Japan Poland Portugal Singapore
South Korea Switzerland Thailand Turkey Ukraine Vietnam

Oxford is a registered trade mark of Oxford University Press
in the UK and in certain other countries

Published in the United States
by Oxford University Press Inc., New York

© Oxford University Press 2005

British Library Cataloguing in Publication Data

Data available

Library of Congress Cataloging-in-Publication Data

Bardgett, Richard D.
   The biology of soil : a community and ecosystem approach / Richard D. Bardgett.
      p. cm.
   Includes bibliographical references and index.
   ISBN 0–19–852503–6 (alk. paper) — ISBN 0–19–852502–8 (alk. paper) 1. Soil
biology. I. Title.
   QH84.8.B35 2005
   577.5'7—dc22   2004030579

Typeset by Newgen Imaging Systems (P) Ltd., Chennai, India
Printed in Great Britain
on acid-free paper by
Biddles, King's Lynn

ISBN 0–19–852502–8 (Hbk) 9780198525028
ISBN 0–19–852503–6 (Pbk) 9780198525035

10 9 8 7 6 5 4 3 2 1

Preface and acknowledgements

For much of history, few things have mattered more to humans than their relations with soil. This is evidenced by a rich historical literature on aspects of soil management and soil fertility, dating back to texts of ancient civilizations of the Middle East, the Mediterranean, China, and India (see McNeill and Winiwarter 2004). Despite the importance of soils to humans, it is really only within the last few decades that ecologists have started to look deeply into the ecological nature of soil habitat, in particular exploring the complex nature of soil biological communities and their environment, and trying to determine the functional significance of soil biota for ecosystem processes. Ecologists are also increasingly becoming aware of the important roles that soil biota and their interactions with plants play in controlling ecosystem structure and function, and in regulating the response of ecosystems to global change. As noted in a recent commentary in the journal *Science* (Sugden et al. 2004), interest in soil ecology is booming, leading to significant advances in understanding of the causes and consequences of soil biological diversity, and of the mutual influences of below-ground and above-ground components of ecosystems.

This increase in interest was the main motivation for this book, to provide students and researchers interested in soil ecology with a comprehensive introduction to what is known about soil biodiversity and the factors that regulate its distribution, and of the functional significance of this below-ground biodiversity for ecosystem form and function. Much is still to be learned about the soil, and this book hopefully highlights some of the many challenges that face ecologists in their exploration of soil. A particular aim of the book is to illustrate how crucial the complexities of the below-ground world are for understanding ecological processes that have traditionally been viewed from a 'black-box' (i.e. its inhabitants grouped as one) or from an entirely above-ground perspective. The book is primarily concerned with biotic interactions in soil and their significance for ecosystem properties and processes. It does not provide a detailed account of the biology of individual organisms present in soil or of the biochemical nature of soil processes. For this, the reader is referred elsewhere.

There are many people that I would like to thank who have helped in the writing of this book. I came to be fascinated by the land, as a child growing up in Cumbria, northern England. This interest was nurtured by my parents, and through the teaching of Eric Rigg, who first introduced me to the scientific discipline of soil. My interest in the land, and soil in particular, further deepened on leaving school, largely through working during a gap year on various aspects of terrestrial ecology. During that time, I worked for Juliet Frankland on the ecology of decomposition, George Handley as a farm labourer, and Carol Marriott on the role of nitrogen fixation in grassland. My tutors at Newcastle University, notably Peter Askew and Roy Montgomery, then deepened my interest in the land further. However, my fascination with soil biology really grew when I worked for Professor Keith Syers, the then Head of the Department of Soil Science at Newcastle. Keith employed me as a research assistant for a short time after completing my degree to explore historical literature on the biological nature of soil fertility. This job set me off on a professional career in the biology of soil. Since that time numerous colleagues have educated and inspired me, providing me with the intellectual resources needed to write this book. In particular, I would like to acknowledge my PhD supervisors, Juliet Frankland and John Whittaker, who introduced me to the complexities of the soil food webs and its role in driving ecosystem processes; James Marsden, of the then Nature Conservancy Council, who deepened my interest and knowledge of the relations between the vegetation and its management; Des Ross and Tom Speir, who taught me how to measure microbial properties of soil, and; Gregor Yeates, Diana Wall, and Roger Cook, who introduced me to the fascinating world of nematodes. In recent years, my own interests have tended to move more above-ground, trying to understand how plant and soil communities interact with one another and how these interactions influence ecosystem processes. Several people have inspired this interest, namely Bob Callow of Manchester University, who taught me how to observe individual plant species in the field, and Lars Walker, Roger Smith, David Wardle, Rene Van der Wal, Wim van der Putten, and John Rodwell who have introduced me to their worlds of plant ecology. It has been a great pleasure to work with all these people.

I am extremely grateful to Ian Sherman of Oxford University Press, who persuaded me to write the book in the first place, and provided much encouragement and advice during its writing. Many colleagues and students have also contributed greatly, providing information and critical comment. In particular, I would like to thank Trevor Piearce who read through the entire manuscript and offered valuable comments, and Roger Cook, David Wardle, Heikki Setälä, Gerhard Kerstiens, Lisa Cole, Kate Carline, Rene van der Wal, Helen Gordon, Edward Ayres, Phil Haygarth, Lars Walker, Helen Quirk, and Ian Hartley who proof read, or offered invaluable advice, on the content of chapters. Edward Ayres compiled Tables 6.1 and 6.2, and several people kindly provided figures and photographs that have been included in the text.

Most of all, I would like to thank my wife, Jill, who gave me much encouragement and support in writing the book, and who tolerated the many weekends, early mornings, and late nights I spent writing. Without her support, and that of my daughters, Alice, Lucy, and Marianne, this book would not have been possible.

Lancaster, November 2004   Richard Bardgett

Contents

1 The soil environment

1.1 Introduction

Soil forms a thin mantle over the Earth's surface and acts as the interface between the atmosphere and lithosphere, the outermost shell of the Earth. It is a multiphase system, consisting of mineral material, plant roots, water and gases, and organic matter at various stages of decay. The soil also provides a medium in which an astounding variety of organisms live. These organisms not only use the soil as a habitat and a source of energy, but also contribute to its formation, strongly influencing the soil's physical and chemical properties and the nature of the vegetation that grows on it. Indeed, along with vegetation, the soil biota is one of five interactive soil-forming factors: parent material, climate, biota, relief, and time (Jenny 1941). The first step towards understanding what controls the abundance and activities of these organisms, and also the factors that lead to spatial and temporal variability in soil biological communities, is to gain an understanding of the physical and chemical nature of the soil matrix in which they live. This chapter provides background on the factors responsible for regulating soil formation, and hence the variety of soils in the landscape. It also discusses the key properties of the soil environment that most influence soil biota, leading to variability in soil biological communities across different spatial and temporal scales.

1.2 Soil formation

In order to understand the properties of soils that influence the biota that dwell therein, we must first consider some of the factors that lead to variations in soils and soil properties within the landscape. One of the most fascinating features of the terrestrial world is the tremendous variety in its landforms, reflecting a diversity of geological processes that have occurred over millions of years; more recent as factors in the variation are biological processes and the influences of man. Similarly, within any landscape there is an incredible range of soils, resulting from almost infinite variation in soil-forming factors. These are highly interactive, in that they all play a part in the development of any particular soil. Combinations of these factors lead to the development of unique soil types, with a relatively predictable series of horizons (layers) that constitute the soil profile (Fig. 1.1). Of greatest interest to the soil ecologist are those horizons that are at, or close to, the soil surface; this is where most microbes and animals live and where most root growth and nutrient recycling occur. These horizons are referred to as the surface organic (O) horizon, which develops when decomposing organic matter accumulates on the soil surface, or the uppermost A horizon, which is composed largely of mineral material but also intermixed with organic matter derived from above. Soil ecologists are also concerned with the plant litter lying directly on the soil surface, deposited during the previous annual cycle of plant growth. This layer, referred to as the L layer, is often overlooked or even discarded in soil sampling regimes, but it is perhaps the most biologically active and functionally important zone of the soil profile.

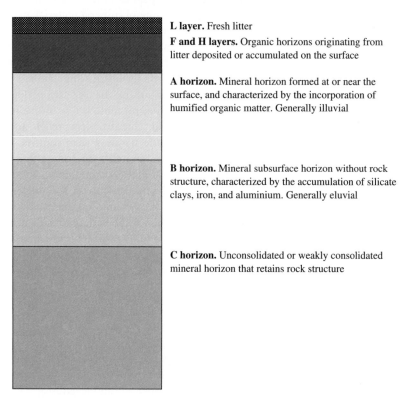

L layer. Fresh litter

F and H layers. Organic horizons originating from litter deposited or accumulated on the surface

A horizon. Mineral horizon formed at or near the surface, and characterized by the incorporation of humified organic matter. Generally illuvial

B horizon. Mineral subsurface horizon without rock structure, characterized by the accumulation of silicate clays, iron, and aluminium. Generally eluvial

C horizon. Unconsolidated or weakly consolidated mineral horizon that retains rock structure

Fig. 1.1 Schematic representation of a soil profile showing major surface and subsurface horizons.

Dense L layer

O horizon

E_a horizon (bleached)

B_s horizon (enriched with Fe)

Fig. 1.2   Podzol (Spodosol in US terminology) soil with deep O horizon (mor) and characteristic bleached E_a horizon above the red, depositional B_s or spodic horizon. (Image by Otto Ehrmann.)

Between this horizon and the A horizon are found layers of organic matter at intermediate stages of decomposition: the F layer, composed of partly decomposed litter from earlier years, and the H layer, made up of well decomposed litter, often mixed with mineral material from below.

While soil profiles vary greatly across landscapes, they can be classified into groups on the basis of their soil properties and soil-forming characteristics, each group having a unique set of ecosystem properties. For example, on free-draining, sandy parent material, in cold and wet climates, usually beneath coniferous forests, podzolic soils develop (Fig. 1.2). These soils, formed by podzolization (Box 1.1), have a deep, acidic surface O horizon, referred to as mor humus. They are subject to heavy leaching and are characterized by low rates of decomposition and plant nutrient availability, and hence low plant productivity. Typically the microbial biomass of these mor soils is dominated by fungi (rather than bacteria) and the fauna are characterized by high numbers of microarthropods (mites and Collembola), and an absence of earthworms. In contrast, on calcium-rich, clayey parent material, typically beneath grasslands and deciduous forests, brown earth soils are often found (Fig. 1.3). These soils have a mull humus composition that is often mildly acidic, owing to leaching of base cations (e.g. calcium) down the soil profile. The mull horizon is characterized by

Box 1.1 Pedogenic processes

Weathering is caused by the action of a range of forces that combine to soften and break up rock into smaller particles that become the parent material of soil. Weathering can occur by physical, chemical, or biological means, and usually by a combination of the three. Physical weathering occurs mainly through the action of water, wind, and changes in temperature, which progressively break down rock into finer particles. Chemical weathering involves the decomposition of minerals by a range of processes including solution, hydration, oxidation, and hydrolysis. Biological weathering occurs under the influence of organisms, for example roots which penetrate and crack open rocks. Organisms such as lichens also produce organic acids that erode the rock surface.

Leaching refers to the downward movement of materials in soil solution, usually from one soil horizon to another. The mobility of elements depends on their solubility in water, the effect of pH on that solubility, and the rate of water percolation through the soil.

Podzolization involves the leaching of Al and Fe from upper soil horizons and their deposition deeper in the soil. These elements are relatively immobile in soil, occurring largely as insoluble hydroxides. Their leaching, however, can be enhanced by the formation of soluble organo-metal complexes, or chelates, which are mobile in percolating water. The most active complexing agents are organic acids and polyphenols, which are especially abundant in the decomposing litter of coniferous trees and ericaceous plants. Removal of Fe from the upper soil horizon leads to the formation of an eluvial E_a horizon, which is bleached in appearance; deposition of Fe deeper in the soil forms a characteristic Bs or spodic horizon, of orange-red colour. The presence of the spodic horizon is a diagnostic feature of the US soil order Spodosol, or a Podzol in UK terminology.

Lessivage refers to the translocation of clay down the soil profile, and its deposition in oriented films (argillans) on ped faces and pore walls. This process gradually forms a subsurface horizon of clay accumulation, which is commonly referred to as an argillic, or B_t, horizon. The end product of this process is the formation of an argillic brown earth (UK soil group), or the US soil order Alfisol. Lessivage occurs under conditions that favour deflocculation of clay minerals. This depends mainly on factors such as clay type and the presence of electrolytes such as Na^+ which cause clay minerals to deflocculate and hence move in water. In acid soils, soluble organic compounds can produce hydrophilic films around clay particles that enhance their movement in percolating water. Clay minerals can also be destabilized by physical impact, for example by raindrops, cultivation, and frost action. Clay movement is unlikely to occur in calcareous soils with high amounts of exchangeable Ca^{2+} ions, which lead to flocculation of clay.

Gleying is the dominant biological process that occurs in hydromorphic soils that are characterized by waterlogging for significant periods of time, owing either to impeded drainage or to a seasonal water table that rises into the subsoil. Gleying is evidenced by the presence of a gleyed horizon that is predominately blue-grey or blue-green in colour. This coloration results from the microbial reduction of ferric (Fe^{3+}) to ferrous (Fe^{2+}) iron that occurs under anaerobic conditions, leading to the solution and depletion of iron from the soil horizon. Mottling is common in gleyed soils owing to re-oxidation and precipitation of Fe in better aerated zones, especially around plant roots and in larger soil pores. This process is evidenced by patches of orange-red colour, which are surrounded by the predominately blue-grey soil matrix. In UK taxonomy, soils are classified as being either ground-water gley soils or surface-water gley soils, the former resulting from a high water table and the latter from impeded drainage.

Fig. 1.3   A typical brown earth soil with mull horizon, under grassland. Note the lack of horizon differentiation caused by intense biological activity and the mixing of organic matter with mineral material from the A horizon (Image by Otto Ehrmann.).

Table 1.1 Soil taxonomy orders

Order	Brief description
Entisols	Recently formed azonal soils with no diagnostic horizons
Vertisols	Soils with swell-shrink clays and high base status
Inceptisols	Slightly developed soils without contrasting horizons
Aridosols	Soils of arid regions
Mollisols	Soils with mull humus
Spodosols	Podzolic soils with iron and humus B horizons
Alfisols	Soils with a clay B horizon and >35% base saturation
Ultisols	Soils with a clay B horizon and <35% base saturation
Oxisols	Sesquioxide-rich, highly weathered soils
Histosols	Organic hydromorphic soils (peats)

intimate mixing of the surface organic and mineral-rich A horizon as a result of the high abundance and activity of soil biota, especially earthworms, leading to high rates of decomposition, nutrient availability, and plant growth. A total of 10 major soil groupings, termed soil orders, have been distinguished by the US Soil Taxonomy (Brady and Weil 1999) (Table 1.1), and 8 major soil groups are recognized by the Soil Survey of England and Wales (Avery 1980); each of these groupings has a unique set of ecosystem characteristics. Further details on these soils and their classification can be found in general soil science textbooks (White 1997; Brady and Weil 1999).

1.3   Soil-forming factors

As noted, within most landscapes there is a tremendous variety of soil types varying in physical and chemical make-up. The soil-forming factors are central to understanding the variability in soils at the landscape level and at the level of the individual soil profile. Being the central forces responsible for creating variety in soil conditions, and hence variations in the habitat of the soil biota, these factors require further consideration. The biota themselves, along with vegetation, constitute one of the main soil-forming factors; both can act as important determinants of soil formation and profile development. This section summarizes some of the important aspects of the main soil-forming factors. It is important to stress, however, that while soil-forming factors are considered individually, they operate interactively in nature, usually with a hierarchy of importance, with one or two of them being pre-eminent in soil development at a particular location.

1.3.1 Parent material

Geological processes acting over millions of years determine the variations and distribution of parent materials from which soils develop. Soils are formed from the weathering of either consolidated rock *in situ* or from unconsolidated deposits—derived from erosion of consolidated rock—that have been transported by water, ice, wind, or gravity. The mineralogical composition of these deposits varies tremendously. For example, the mineralogy of igneous rocks, formed by solidification of molten magma in, or on, the Earth's crust, ranges from base-rich basalts (basic lava) with high amounts of calcium (Ca) and magnesium (Mg) to acidic rhyolites (acid lava) which contain high amounts of silica (Si) and low amounts of Ca and Mg. Rocks of intermediate base status, such as andesites, also commonly occur. Parent material also determines grain size, which determines soil texture (relative proportions of sand, silt, and clay), which in turn affects many soil properties, such as the ability of the soil to retain cations (its cation exchange capacity), the moisture retaining capacity, and soil profile drainage. Such variation in the mineralogy of rocks, therefore, strongly influences the type of soils that are formed and the character of the vegetation that they support (Fig. 1.4). Soils formed from weathering of basic lava, for example, tend to be rich in minerals such as Ca, Mg, and potassium (K) and fine textured (clayey), and have a high ability to retain cations of importance to plant nutrition (e.g. NH_4^+, Ca_2^+). These soils are typically fertile brown earths with biologically active mull humus. In contrast, soils that are formed from acidic lava, such as granites and rhyolites, are low in Ca and Mg, coarse textured (sandy), and hence freely

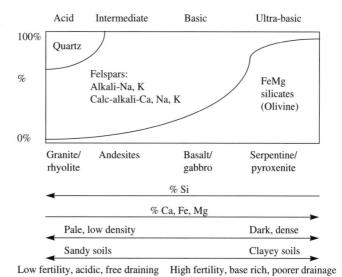

Fig. 1.4   Schematic classification of igneous rocks and their resulting soils.

drained, with low cation retention capacity. The soils that typically develop here are therefore strongly leached, nutrient-poor, acidic podzols with mor humus.

1.3.2 Climate

Historically, climate has been considered pre-eminent in soil formation, owing largely to the striking associations that exist, on continental scales, among regional climate, vegetation type, and associated soils. Indeed, these broad climatic associations led to the development in Russia of one of the first soil classification systems—the zonal concept of soils (Dokuchaev 1879). This system identified so-called zonal soils—those that are influenced over time more by regional climate than by any other soil-forming factor. While climate may play a crucial role in soil development on continental scales, for example, across Russia and Australia, it is arguably not as important in areas such as the subtropics and tropics where land surfaces are much older and more eroded, or in younger landscapes such as Britain where most soils are developed on recent, glacial deposits. Here, the factors of topography and parent material are of greater importance.

The effects of climate on soil development are largely due to temperature and precipitation, which vary considerably across climatic zones. These factors strongly govern the rates of chemical reactions and the growth and activities of biota in soil, which in turn affect the soil-forming processes of mineral weathering and decomposition of organic material. The effects of temperature on soil biological activity are well known; it is generally accepted that there is an approximate doubling of microbial activity and enzyme-catalysed reaction rates in soil for each $10\,°C$ rise in temperature, up to around $30–35\,°C$. Above this temperature, however, most enzyme-catalysed reactions decline markedly, as proteins and membranes become denatured. Some microbes can live at extreme temperatures; for example, cold-tolerant fungi occur in polar soils and remain physiologically active down to $-7\,°C$ (Robinson and Wookey 1997). These cold-tolerant microbes are called pychrophiles, whereas microbes that live in extremely high temperatures are called thermophiles.

The effects of temperature and precipitation on soil formation are especially marked at high altitudes and latitudes. For example, soil organic matter content is often found to increase with increasing elevation, commonly reaching a peak in montane forests (Körner 1999). (At higher altitudes, above the treeline, the organic matter content of soils declines and reaches almost zero in unvegetated substrates in the upper alpine zone.) This increase in soil organic matter content is largely due to declines in temperature and high precipitation, which reduce microbial activity and rates of decomposition. Similarly, in high-latitude regions of Europe, vast peatlands have developed in areas where the combined effects of high

Fig. 1.5   Blanket peat at Moor House National Nature Reserve in northern England. Here, deep peats have developed at high altitudes where high rainfall and low evapotranspiration combine to cause excessive soil wetness that retards decomposition. These peatlands are of special significance because they represent a significant (ca. 30%) store of global terrestrial C. Indeed, the majority of the UK's terrestrial C is stored in the peat soils of northern Britain (Image by Richard Bardgett.).

rainfall and low temperature, and minimal evapotranspiration, have led to anaerobic conditions (waterlogging) and the retardation of decomposition of organic matter. The consequence of this has been the accumulation of great masses of peat (blanket peats), especially in topographically uniform areas where drainage is reduced (Fig. 1.5). Dramatic changes in the physiology and productivity of dominant plants also occur along altitudinal and latitudinal gradients (Díaz et al. 1998), altering the nutritional quality (e.g. N content) of the leaf litter that is produced annually. As will be discussed in Chapter 4, such changes in organic matter quality resulting from shifts in plant community composition can have profound effects on the decomposability of organic matter, and hence the accumulation of organic matter on the soil surface.

1.3.3 Topography

Variations in topography influence soil development, largely through effects on soil drainage and erosion. Soil drainage is primarily affected by the position of a soil on a slope; soils at or near the top of a slope tend to be freely drained with a water table at some depth, whereas those at or near the bottom of the valley tend to be poorly drained with a water table close to the soil surface. These differences in drainage strongly influence soil development, leading to the development of a hydrological sequence (Fig. 1.6): well-drained soils on hilltops have deep, orange-brown subsurface horizons, indicative of oxidation processes (iron in ferric state); as drainage deteriorates towards the valley bottom, the soil profile becomes increasingly anaerobic and blue-grey in colour, indicative of a dominance of reduction processes

(a)

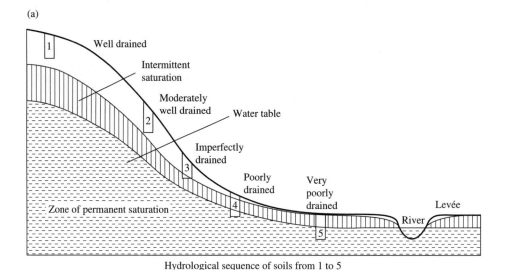

Hydrological sequence of soils from 1 to 5

(b)

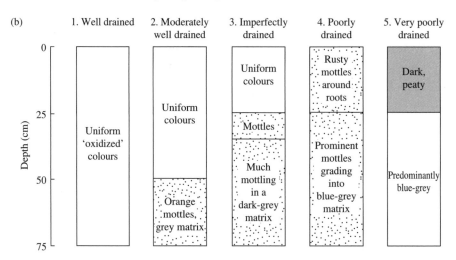

Fig. 1.6   (a) Section of a slope and valley bottom showing a hydrological soil sequence, and (b) changes in soil profile morphology. (Redrawn with permission from Blackwell Science; White 1997)

(iron in ferrous state) and the process of gleying (Box 1.1). In extremely wet valley bottom soils, deep O horizons develop on the soil surface as a result of retardation of decomposition processes under anaerobic conditions.

Slope characteristics also greatly influence soil erosion processes, which in turn affect soil formation. In general, soils on ridges and steeper parts of slopes are shallower than those on lower slopes and valley bottoms, owing to the movement of soil particles down-slope by wash and soil creep. Because erosion preferentially moves finer particles down-slope, the soils of lower slopes and valley bottoms also differ in their mineralogical composition, being more fine textured. In humid regions, soils of lower slopes and valley bottoms also tend to be enriched in base cations and salts, owing to seepage of solutes from higher slopes and hilltops. Soil movement down-slope also leads to the formation of distinct morphological features on slopes, such as terraces. These features are especially common in alpine regions where steep slopes and freeze–thaw motion lead to instability of the surface soil and consequent down-slope creeping. This process is called solifluction. What happens here is that soils become saturated with water and freeze, and then melt; the expansion associated with freezing makes the surface soil very unstable when it thaws. This leads to downward movement of soil even on the gentlest slopes, especially if the subsurface soil is frozen.

1.3.4 Time

The age of the soil is a major factor underlying variations in soils and ecosystem properties. Soils become increasingly weathered over time, and, consequently, soil profiles generally become more differentiated, with more abundant and thicker horizons. This weathering process involves progressive leaching downwards of elements and minerals in percolating water. In particular, during the process of podzolization, progressive downward movement of iron (Fe) and aluminium (Al) leads to the development of a spodic (Box 1.1) subsurface horizon that is enriched with these minerals, and an upper E_a horizon, whence these elements have been removed, that is bleached in appearance. Over time, clay minerals also become leached down-profile, a process called lessivage (Box 1.1), leading to the development of subsurface argillic horizons. Other important changes in soils that occur with age are increases in soil organic matter and nitrogen (N) content (Crocker and Major 1955) and, over very long timescales (hundreds of thousands or millions of years), a progressive reduction in the availability of soil phosphorus (P) owing to its loss from the system and fixation in mineral forms that are not available to plants (Walker and Syers 1976).

The relationship between time and soil development is best illustrated by examining soil chronosequences, which are places where, for various reasons, a sequence of differently aged, but otherwise similar, geologic substrates exists. Glacier Bay, on the coast of southeast Alaska, is one of the best known places for research on soil and ecosystem development because of the continuous retreat of the glaciers since 1794. Furthermore, records of the retreat, over some 100 km, have been maintained since this time, so the age of the glacial moraines is known. This has resulted in a site chronology over a period of 200 years, and from this, patterns of soil development can be tracked along with the succession in vegetation from the initial pioneer plant communities on recent moraine through to the climax spruce forest on the oldest moraines (Box 1.2). This chronosequence represents stages in the development of a podzol; as time progresses, the soil profile becomes progressively deeper and differentiated, the oldest soils being acidic in nature and having a thick organic surface horizon, a thin bleached horizon, and subsurface spodic horizon (Crocker and Major 1955). As organic matter steadily accumulates in the surface organic horizon, the amount of N in soil also increases; the organic carbon (C) and N content of underlying mineral soil also builds up (Fig. 1.7). Of particular significance for organic matter and N accumulation is the early stage of vigorous alder growth on terrain that has been ice-free for some 75 years. Here, N accumulates at a rate of about 4.9 g N m^{-2} yr^{-1}, reaching values of some 250 g N m^{-2} within around 50 years of soil development (Crocker and Major 1955). Although there is no single explanation for the soil development sequence at Glacier Bay, it appears that establishment of plants and intensive leaching have played key roles (Matthews 1992).

Studies at Glacier Bay demonstrate progressive soil development over relatively short timescales towards climax, while other sites can be used to demonstrate soil change over hundreds of thousands, or even millions of years. The Hawaiian island archipelago, for example, presents a chronology of soil development over some 4.5 million years; Kilauea volcano on the Island of Hawaii is active, while Kauai, the oldest site, on the northwest end of the high islands is estimated to be 4.5 million years old (Fig. 1.8). A range of intermediate aged sites is also present, and all sites are derived from volcanic lava of similar mineralogy and have vegetation that is dominated by the same tree, *Metrosideros polymorpha* (Fig. 1.9). A key feature of this chronology is that it demonstrates a reduction in P availability in the oldest sites, which causes a dramatic decline in plant productivity (Crews et al. 1995). This fall in P availability follows the theoretical model of soil development that was proposed by Walker and Syers (1976): as soils age, P becomes surrounded, or occluded, by Fe and Al hydrous oxides, rendering the P largely unavailable to plants and also the soil biota (Fig. 1.10). This process of occlusion is especially likely to occur in very old soils because prolonged weathering of minerals leads to the formation of Fe and Al oxides that have a strong affinity for P. This P limitation to vegetation is further exacerbated in old soils because low soil fertility sets in motion a feedback whereby reductions in biological activity in soil reduce decomposition of plant litter, further intensifying nutrient limitation

Box 1.2 Glacial moraine succession at Glacier Bay, southeast Alaska

Many glaciers in the Northern Hemisphere have been retreating during the past 200 years. Glacier Bay, on the coast of southeast Alaska, is one of the best known places for research on primary succession because of the continuous retreat of the glaciers since 1794. Furthermore, records of the retreat, over some 100 km, have been kept since this time, so the age of the glacial moraines is known. This produces what is known as a site **chronology** over a period of 200 years. The map of the Glacier Bay fjord complex shows the rate of ice retreat since 1760. The dashed lines show the approximate edge of the ice in 1760 and in 1860.

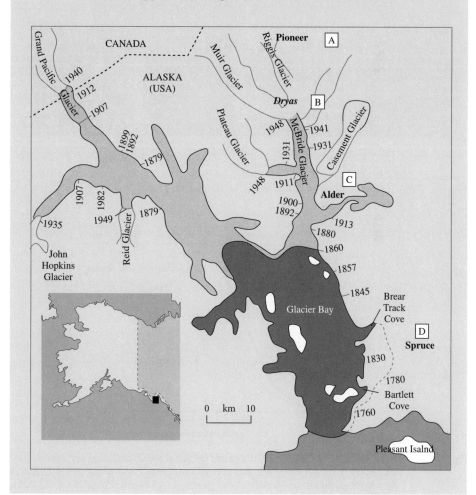

At Glacier Bay, there are four stages of succession:

1. The **pioneer stage** has been ice-free for 10–20 years. The ground is mainly bare glacial till, but there are patches where the till has been colonized by algal crusts, lichens, mosses, *Dryas*, and scattered willows.
2. The ***Dryas* stage** is on ground that has been ice-free for some 30 years. Here, the ground surface is an almost continuous mat of *Dryas* with scattered willow, cottonwood, and alder.
3. The **alder stage** forms after about 50–70 years. Here the *Dryas* is taken over by dense thickets of the nitrogen-fixing plant alder.
4. The **spruce stage** forms after 100–150 years of succession. Here, spruce trees that have replaced the alder dominate the canopy.

(1)   (2)   (3)   (4)

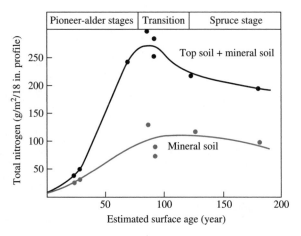

Fig. 1.7   Total N content of soils recently uncovered by glacial retreat at Glacier Bay, Alaska. Plant succession is shown along the top. (Data from Crocker and Major 1955) Shillawait permission–could not say (Reprinted with permission of Pearson Education; Kerbs 2001, using data from Crocker and Major 1955)

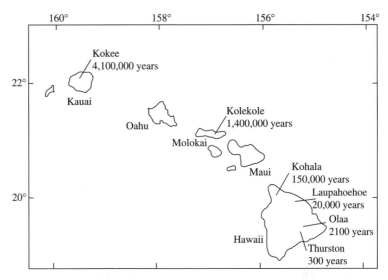

Fig. 1.8   The Hawaiian Island archipelago presents a chronology of soil development of some 4.5 million years: Kilauea volcano on the Island of Hawaii is active, while Kauai, the oldest site, on the northwest end of the high islands is estimated to be 4.5 million years old. (Redrawn from Crews et al. 1995).

Fig. 1.9   A *Metrosideros polymorpha* ('ohi'a lehua) forest on the Hawaiian island of Molokai with low stature and biomass. This ecosystem developed in the absence of catastrophic disturbances during the past 1.4 million years. Depletion of P at this site has led to a decline in forest productivity. (Image by Richard Bardgett.)

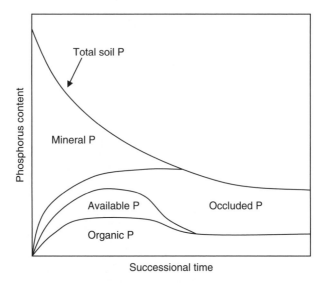

Fig. 1.10   Generalized effects of long-term weathering and soil development on the distribution and availability of P in soil (Adapted from Walker and Syers 1976).

(Crews et al. 1995). Another excellent example of P limitation in very old sites is the large sand dune systems on the subtropical coast of Queensland, Australia. These dune systems provide a chronology of soil development, without interruption from glaciation, dating back some 400,000 years (Thompson 1981). Here, progressive weathering and leaching in the free-draining sand over this period has led to the development of so-called giant podzols (Thompson 1981) with soil profiles of some 20 m depth, which are characterized by extreme P limitation and stunted tree growth (Fig. 1.11). A detailed account of this soil development sequence can be found in Thompson (1992).

1.3.5   Human influences

An increasing proportion of the Earth's surface is under some form of management by man, and human impacts on ecosystems are significant and growing (Vitousek et al. 1997). The most obvious effects of humans on soil development result from the widespread destruction of natural vegetation for agricultural use, for example, in the tropics where some 8% of rainforest is thought to be cut down per decade (Watson 1999). Human activities also have dramatic effects on soil development by directly modifying the chemical and physical nature of the soil environment by fertilization, irrigation, and drainage, and by ploughing for cultivation. Humans can also have important indirect effects on soil development. For example, increasing concentrations of CO_2 in the atmosphere have an effect on vegetation,

Fig. 1.11   Stunted tree-growth resulting from P limitation in old-growth (400,000-year-old) eucalyptus forest on the subtropical coast of Queensland, Australia. (Image by Richard Bardgett.)

which in turn affects soil processes and soil biota. The introduction of invasive species into natural ecosystems can also have profound effects on soil properties and biota, for example, by changing the quality of litter inputs into soil and/or soil nutrient availability. Effects of human activities on soil biota and soil properties will be discussed in detail in Chapter 6.

1.4   Soil properties

Variation in soil-forming factors determine the physical and chemical nature of soils, which in turn influences greatly the nature of the soil biota and hence ecosystem properties of decomposition and nutrient cycling. Variation in soil properties, especially the physical matrix of the soil, also greatly influences the movement of water and associated materials both within and between ecosystems. This section examines some of the key soil properties that most strongly influence the soil biota and their activities, and the nature of ecosystems.

1.4.1 Soil texture and structure

The term soil texture refers to the relative proportions of various-sized particles—sand (0.05–2.0 mm), silt (0.002–0.05 mm), and clay (<0.002 mm)—within the soil matrix (Fig. 1.12). It is primarily dependent on the parent material from which the soil is formed and the rate at which it is weathered, as discussed above. Soil texture is of importance largely because it determines the ability of the soil to retain water and nutrients: clay minerals have a higher surface area to volume ratio than sand and silt, and hence soils with a high clay content are better able to hold water by adsorption and to retain cations on their charged surfaces. The ability of a soil to retain cations (e.g. Ca^{2+}, Mg^{2+}, NH_4^+) is referred to as its cation exchange capacity, which reflects the capacity of clay minerals to hold cations on negatively charged surfaces. This retention of cations on clay minerals represents a major short-term store of nutrients for plant and microbial uptake.

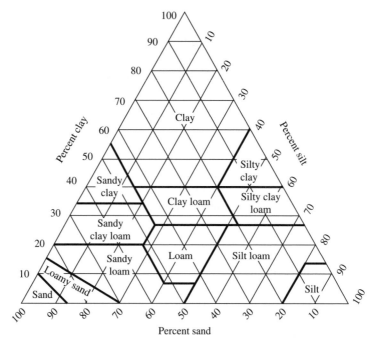

Fig. 1.12   Composition of the textural classes of soils based on percentages of sand, silt, and clay. For example, a soil with 60% sand, 10% silt, and 30% clay is a sandy clay loam.

Soil structure reflects the binding of the above various-sized mineral particles into larger aggregates or a ped. The actual formation of stable aggregates requires the action of physical, chemical, and biological factors: (1) freeze–thaw and soil shrinkage and swelling help to mold the soil into aggregates; (2) the mechanical impact of rain and ploughing reorganizes soil materials; (3) the activities of burrowing animals, such as earthworms, lead to the mixing of mineral and organic materials and the formation of stable organo-mineral complexes; (4) the faeces of soil animals can act as nuclei for aggregate formation; (5) fine roots and microbes produce a range of polysaccharide glues which bind soil particles together, and fungal hyphae literally hold together mineral particles and organic matter. Together, these factors combine to produce stable aggregates in soil. As will be discussed in Chapter 3, large soil animals, such as earthworms and termites, can also substantially affect soil structure by creating macropores and channels as a consequence of their feeding and burrowing activities; this in turn increases infiltration rates, and hence drainage of water through soil.

Soil aggregation and structure is of concern to the soil ecologist, not only because the activity of the soil biota strongly affects it, but also because the structure of soil determines the physical nature of the living space. Aggregation determines the pore distribution of soil, which affects both the distribution of water in soil (specifically the degree to which pores are filled with water) and the extent to which biota are able to enter and occupy pore space, which is controlled by pore neck diameter and the size of the organism. For example, nematodes are approximately 30 μm in diameter, so their migration in soil is restricted by pores of diameter <30 μm and is optimal in soils with particle sizes in the region of 150–250 μm. Fungi and bacteria are much smaller in size, so they can occupy much smaller pores (1–6 μm) than can nematodes. Because the primary food source of many nematodes is microbes, these smaller pores can therefore provide a refuge for microbes from their predators, thereby affecting the nature and intensity of trophic interactions in soil.

In sum, good soil structure is recognized as a key attribute of fertile and biologically active soil, because it increases the flow of water and gases through soil, reducing the possibility of the development of anaerobic conditions, which would be detrimental to soil biota and their activities, and harmful to plant growth. Good soil structure promotes free movement of biota, thus increasing opportunities for trophic interactions; it allows roots to proliferate and enables aerobic microbial processes to dominate.

1.4.2 Soil organic matter

The organic matter content of soil varies tremendously in terms of both its chemical composition and its quantity. As already discussed, this depends on a variety of interacting factors such as vegetation type, climate, parent material, soil drainage, and the activity of soil biota: a particular combination of these factors generally leads to the formation of either mull or mor humus forms. Soil organic matter is of importance because it promotes soil structural stability (see above), thereby preventing soil erosion. It is also extremely effective in retaining water within the soil matrix. Soil organic matter is of particular importance for biota because it is their primary source of nutrients and C. Through a range of activities, soil biota decompose organic matter, converting it into simple organic molecules that they can assimilate and use for energy and growth. Most components of soil organic matter occur as large molecules (e.g. cellulose, amino sugars, proteins, nucleic acid, lignin, etc.), which, over time, are depolymerized by extracellular microbial enzymes to yield simpler units which can then be assimilated by the microbes for energy and C. For example, the enzyme cellulase breaks down cellulose—the principal component of plant tissue—into glucose units that are readily assimilated by microbes and used for energy and C. Microbial enzymes also produce soluble organic nutrients (e.g. amino acids), which are either absorbed by microbes to meet their nutrient requirement (e.g. for C and N) or, in some circumstances, taken up directly by plants and mycorrhizal fungi (Chapter 3). Soluble organic nutrients that are absorbed by microbes are either retained to meet their C and N needs, or, when they require only C to support their energy needs, the excess N is secreted into the soil environment as inorganic N (ammonium), which is then available for plant uptake. This process is called mineralization; it is of key importance at the ecosystem scale because it determines the availability of inorganic nutrients for plant uptake, and hence plant productivity (Chapter 3). Soil fauna also contribute to the decomposition process and nutrient mineralization both directly, by mixing and fragmenting organic matter into smaller units that are more accessible to microbial attack, and indirectly, by feeding on microbes, affecting their growth and activity, and excreting nutrients that are in excess of the animals' requirements into the soil environment (Chapter 3).

The rate at which organic inputs to soil are decomposed depends primarily on their quality, which is dependent on the type of compounds that are present within them. Rapidly decomposing materials, such as the litter of deciduous trees and animal faeces, generally contain high amounts of labile substances, such as amino acids and sugars, and low concentrations of recalcitrant compounds such as lignin. In contrast, the litter of coniferous trees decomposes slowly, being rich in large, complex structural compounds such as lignin and defence compounds such as polyphenols; this material is also unpalatable to soil fauna, further slowing down its decomposition. The importance of variation in the rate of decomposition at the ecosystem scale relates to the production of CO_2 from heterotrophic microbial activity and its evolution into the atmosphere, and to the conversion of organic nutrient forms to simple inorganic nutrients (e.g. ammonium and phosphate) that are available for plant uptake, a strong determinant of plant productivity. These factors will be discussed in detail in later chapters.

1.4.3   Soil water

It is water that renders the soil environment habitable. Together with dissolved nutrients it makes up the soil solution, an important medium for supplying nutrients and water to growing plants. It also provides a medium for many soil biota to live and move around in. Nematodes and protozoa, for example, live in water films and in free water, along with the bacteria that they feed upon. Larger fauna, such as microarthropods (mites and Collembola), live in open pore spaces but are very sensitive to desiccation, migrating out of soil when it becomes dry.

There are certain water-holding characteristics of soils that determine the amount of water that is available to plants and soil biota. Under normal conditions, soil pores will contain air as well as water, and most water is held in pore spaces as films or as water absorbed onto soil particles. Under these conditions the soil is said to be unsaturated. After a period of heavy rain or irrigation, however, pore spaces in soil become filled with water and the soil becomes saturated. After a period of drainage, when the amount of water held in the soil is in equilibrium with gravitational suction (5–10 kPa), field capacity is reached and no more water drains from the soil. The pores that are drained at field capacity are the macropores, which are those that are created by the burrowing activities of animals and by plant roots. As plant roots continue to absorb water from the soil, the amount of water held in films is reduced, and the remaining water adheres tightly to soil particles. At this stage, water moves through thin films to plant roots in response to a gradient of water potential; if water is not replenished and plants continue to take it up, a situation is reached when plants can no longer remove water that is strongly adhered to particle surfaces. This is the permanent wilting point, and the difference between this measure and field capacity is referred to as the water-holding capacity of the soil. In general, the water-holding capacity is greater in soils that contain large amounts of clay and/or organic matter, because these components have high surface areas that readily retain water. Also, clay soils have more small pores that readily retain water under gravitational suction than do sandy soils, which have larger pores that are more easily drained.

1.4.4   Soil pH

A large proportion of the Earth's soils are acidic, especially in the tropics, where ecosystems persist at soil pH values of 4 or less (pH here is a measure of the concentration of H^+ ions in soil water). Many northern ecosystems also have very acidic soils: the pH values of the soils of Boreal forests and heathlands are often of 4 or less. In many parts of the world, soil acidity is further exacerbated by the use of inorganic fertilizers and acid rain (Kennedy 1992). Soils become acidic if base cations (e.g. Ca^{2+}, Mg^{2+}, K^+) are leached from the soil profile, to be replaced by H^+ and Al^{3+} ions on cation exchange sites. Acidity in soils can come from various sources: (1) carbonic acid, which is formed by the dissolution of CO_2 in water, dissociates to yield H^+ ions; (2) microbial oxidation of ammonium ions (NH_4^+) to nitrate (NO_3^-), the former being derived from mineralization and fertilizer inputs, also yields H^+ ions; (3) atmospheric pollution (acid rain) and natural sources of acids, including volcanic eruptions and thunderstorms that yield sulphur dioxide and oxides of N, respectively, produce sulphuric and nitric acids that acidify soils (it has been proposed that the widespread occurrence of acidic soils in tropical and subtropical regions is, in part, a result of high thunderstorm activity in these regions. Long-term weathering and leaching of cations also contributes significantly to acidity in these old soils); and (4) decomposition of organic matter that has high concentrations of phenolic and carboxyl groups liberates H^+ ions.

Soil pH is of concern to the soil ecologist because it controls nutrient availability and it directly impacts on soil biota. Acidic soils, for example, are characteristically high in soluble aluminium (Al^{3+}), which can be toxic to plants and microbes. This is because aluminium occurs largely as insoluble forms, for example, as part of clay minerals whose structure becomes unstable at low pHs (4–5), releasing aluminium ions into soil solution (Kennedy 1992). The P availability in soil is typically low under acidic conditions, owing to the formation of iron and aluminium phosphates. These phosphates dissolve to release P into soil solution as pH rises, making it available for plant uptake; the availability of P is typically greatest between pH 6 and 7 (Chapter 3).

The effects of pH on soil organisms are well documented, and approximate tolerances of major groups of soil organisms are known. Most microbes grow within the pH range 4–9, although acidophiles are known to survive at pHs as low as 1, for example, sulphate-oxidizing bacteria (species of the genus *Thiobacillus*) which occur in hot springs and mine wastes. Growth in extremely alkaline conditions is restricted to a few fungi and bacteria, but such conditions tend to occur only in soda lakes and deserts, for example, those in Egypt which have pH values between 9 and 11. Soil animals are also very sensitive to acidic conditions in soil. For example, earthworms occur in very low numbers, with few species, in most acidic soils, and they become progressively more abundant as soil pH increases to neutrality (Edwards and Bohlen 1996). Acid soils tend to be dominated by enchytraeid worms (potworms), reaching densities as high as $100 \times 10^3 \, \text{m}^{-2}$ in acid peat soils (Cole et al. 2002*a*). Under such conditions, these enchytraeid worms replace earthworms as the functionally dominant soil animal.

1.5 Conclusions

Within most landscapes, there is a tremendous variety of soil types, reflecting spatial variations in the operation of a range of interacting soil-forming factors and pedogenic process, and the influences of man. The need for the soil ecologist to recognize this variation, at both spatial and temporal scales, cannot be stressed enough because it means that the roles of biota, relative to other factors, in controlling ecosystem processes (e.g. nutrient cycling, hydrological fluxes, and plant productivity) and soil formation itself will be context dependent. In other words, the roles of biota and the factors that regulate their community structure and activities will vary from soil to soil, depending on the dominant physical and chemical characteristics of that soil. In view of this, it is essential that studies on the soil biota are accompanied by detailed characterization of the physical and chemical nature of their habitat, and an appreciation of the soil-forming factors that led to the development of that soil in the first place.

2 The diversity of life in soil

2.1 Introduction

Ecologists have long been fascinated by the vast diversity of organisms that live on Earth. Indeed, understanding the complex patterns of this diversity, and the dominant forces that control them, has been a major focus of community ecology. This area of ecology has, however, had an almost exclusively above-ground focus (Mittelbach et al. 2001), with very little effort being put into characterizing and understanding the significance of below-ground diversity. As a result, information on the actual diversity of groups of soil biota is very sparse compared with what is known about above-ground organisms, especially at the species level. Furthermore, the information that is available is restricted largely to a few ecosystem types and a few taxonomic groups of organisms within them. This relative lack of attention to soil biota is surprising, considering that the majority of the Earth's species might actually be living in soil (Wardle 2002). This inattention is, however, understandable; soil organisms are not easily seen, they are extremely difficult to study, and they lack the sentimental appeal that many above-ground species attract.

Despite all this, an increasing number of ecologists are starting to turn their attention to soil communities, largely because of an awareness that not only do soil organisms regulate major ecosystem processes, such as organic matter turnover and nutrient cycling, but they also act as important drivers of vegetation change. The aim of this chapter is to provide a brief introduction to some of the major groups of organisms that live in soil and to discuss their diversity in different ecosystems. The chapter will also consider what factors are likely to act as primary determinants of soil biodiversity across various spatial and temporal scales.

2.2   The soil biota

The vast array of microbes and animals that live in soil constitutes the soil food web, whose primary role in ecosystems is the recycling of organic matter from the above-ground plant-based food web. The most numerous and diverse members of the soil food web are microbes—the bacteria and fungi—but there are also many animal species of varying sizes that live in soil, including the microfauna (body width <0.1 mm; e.g. protozoa and nematodes), the mesofauna (body width 0.1–2.0 mm; e.g. microarthropods and enchytraeids), and the macrofauna (body width >2 mm; e.g. earthworms, termites, and millipedes) (Fig. 2.1). These fauna cross a range of trophic levels, and in soil food webs they are often allocated to functional

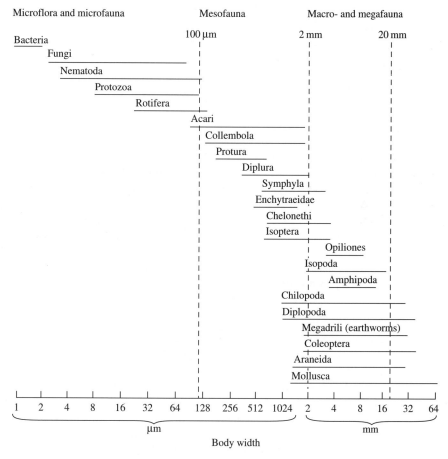

Fig. 2.1   Classification of soil biota on the basis of their body size. (Adapted from Swift et al. 1979).

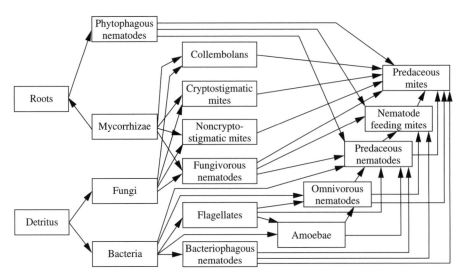

Fig. 2.2   Structure of the soil food web. (Adapted from de Ruiter et al. 1995)

groups based on their feeding habit (Fig. 2.2). Some feed primarily on microbes (microbial-feeders) or litter (detritivores), whereas others feed principally on plant roots (herbivores) or other animals (carnivores). Recent studies reveal that omnivores are also very common in soil food webs, in that many soil animals appear to feed across different trophic levels (Ponsard and Arditi 2000; Scheu and Falca 2000). Detailed descriptions of the major components of the soil food web are beyond the scope of this book. Here, the main players are briefly introduced; the reader is refereed to the *Encyclopaedia of Soil Science* (Lal 2002) for more detailed accounts of soil fauna, and to *Soil Microbiology and Biochemistry* by Paul and Clark (1996) for information on microbes.

2.2.1 The primary consumers

The primary consumers of the soil food web are the microbes (bacteria, fungi, actinomycetes, and algae) that are primarily responsible for breaking down and mineralizing complex organic substances. Microbes are by far the most numerically abundant and diverse members of the soil food web, with literally thousands of microbial species being present in soil. The two most abundant groups of microbes are the fungi and bacteria. Fungi are distinct from bacteria in that they are eukaryotic and generally produce filamentous hyphae that can penetrate and explore microhabitats of the soil. In contrast, bacteria are prokaryotic and unicellular. For mobility, bacteria rely on the presence of flagella that enable them to move through water films, or if flagella are absent, they rely on passive transport through soil via roots, fauna, or the general movement of water through soil. Until recently, it was impossible to detect the majority of microbes in soil, since most are uncultivable using traditional culturing techniques. The recent development of culture-independent molecular tools has allowed ecologists to start to explore the true diversity of bacteria and fungi in soils and other habitats (O'Donnell et al. 2005; Robinson et al. 2005) (Box 2.1).

Box 2.1 Methods for characterizing microbial communities in soil

Until recently, progress in soil microbial ecology was hampered by the inadequacy of methods to characterize the vast diversity of microbial communities in soil. This is due largely to the non-culturability of most microbial cells—for example, in soil, the portion which can be cultured in the laboratory has been estimated to be 0.3% of the total number of cells observed microscopically—and also due to problems involved in extracting micoorganisms from complex and variable matrices. In recent years, a battery of molecular methods have been developed that enable the true diversity of microbial communities to be assessed, since they do not rely on the cultivation of microorganisms prior to analysis.

Phenotypic analysis. The direct derivation of cellular components from entire microbial communities represents one important method for quantitative assessment of microbial abundance and composition. The most commonly used approach for this is Phospholipid Fatty Acid Analysis (PLFA). The PLFA technique relies on the fact that phospholipids exist in the membranes of all living cells and that different subsets of the microbial community contain different PLFAs, or at least differ in their fatty acid composition (Tunlid and White 1992). Furthermore, PLFA measures only the living part of the microflora; phospholipids are not used as storage material, and when released upon cell death they are used as substrate by living microorganisms, and within minutes to hours are metabolized to diglyceride and PO_4^{3-} (White et al. 1979). The assessment of PLFA pattern of an environmental sample, therefore, can be viewed as an integrated measurement of all living microorganisms present in that sample, providing a quantitative measure of the different microbial groups within mixed microbial communities. However, it should be emphasized that since most PLFAs exist in different concentrations in a taxonomically wide range of microorganisms (Ratledge and Wilkinson 1988), it cannot be used to measure specific species or genera within entire microbial communities; it can be used only to determine changes in the relative abundance of very broad taxonomic groups.

Genotypic analysis. Over the past decades, molecular biological techniques based on analysis of the heterogeneity of RNA and DNA sequences have revised microbial taxonomy and greatly expanded our knowledge of the microbial ecology of soil. These techniques have been made possible by our ability to extract and purify DNA and RNA from soil. Genes providing information on the presence and diversity of a single physiological group (via a functional gene) or indicating the phylogenetic relationship of detected microorganisms with other organisms (via universally distributed, conserved genes) have been used. Among the most widely used phylogenetic genes are those of the RNA of the small subunit of the ribosome, the 16S rRNA and 18S rRNA genes of prokaryotic and eukaryotic microorganisms, respectively. This is due, in part, to the accepted use of a phylogenetic system based on rRNA sequences, where the extent of similarity between ribosomal sequences provides an estimate of phylogenetic or evolutionary relationships (Woese 1987). The 16S rRNA gene, and homologous 18S rRNA gene, contain conserved regions, present in all microbes, and variable regions that can be used to distinguish between sequences from different species or strains. The rRNA sequence database was initially compiled from well characterized strains from culture collections (Larsen et al. 1993) and now represents a dynamic and ever increasing dataset against which new sequences may be compared. The database includes sequences from cultivated strains as well as from cloned gene sequences from genetic material extracted directly from environmental samples (Pace et al. 1986). The genetic diversity of soil bacterial communities has been studied by a variety of nucleic acid hybridization techniques; including gene probe analysis and restriction enzyme analysis. Development of the polymerase chain reaction (PCR) has revolutionized studies of microbial ecology and led to new methods for the analysis of bacterial community structure and diversity in soil samples. Since the mid-1980s, the majority of PCR-based studies of microbial ecology have centred on the analysis of bacterial 16S rRNA. More recently, the 18S rRNA molecule of fungi has also been used as a target in ecological studies (Kowalchuck et al. 1997; Anderson and Kohn 1998). In addition to rRNA-based analyses, the detection and analysis of specific functional genes can also be used to provide useful information on the microbial ecology of environmentally important processes.

The primary role of fungi is to decompose organic materials via the production of a vast array of extracellular enzymes. By virtue of their large biomass (which can often exceed that of other microbes, plants, and animals) they represent a significant portion of the ecosystem nutrient pool. In grassland, for example, there can be as much as 250 kg ha^{-1} dry

Table 2.1 Comparison of the distribution of the biomass of basidiomycetes (kg ha^{-1} dry wt) with that of other microbial decomposers in the floor of a temperate deciduous woodland with mull humus (Meathop Wood, Cumbria, UK; Frankland, 1982). Mycelial biomass was estimated from measurements of fungal hyphal length by direct microscopic observation

Substrate or soil horizon	Basidiomycetes		Other fungi		Bacteria and actinomycetes	
	Living	Total	Living	Total	Living	Total
Woody debris	30.5	216.9	7.3	34.7	2.6	601.6
L	3.1	8.7	0.5	4.1	37.3[a]	8433.3[a]
(Oh and Ah)	8.9	31.7	3.4	12.9		
A	<1.0	<1.0	26.4	97.5		
B	<1.0	<1.0	31.4	155.6		
Dead roots	228.0	1625.5	65.1	325.7	8.0	1851.1
Total	271.5	1886.8	134.1	630.5	47.9	10,866.0

[a] Values are the sum of all horizons.

fungal hyphae within the surface 5 cm soil (equivalent to over 3 km hyphae g^{-1} dry soil), containing as much as 8 kg of N and 2 kg of P (Bardgett et al. 1993a). Some 10 times as much fungal biomass is found in deciduous woodland soils, where the soil surface has been referred to as being 'packed' with fungal mycelium (Frankland 1982) (Table 2.1). As well as degrading relatively permanent pools of organic matter in soil, these extensive mycelial networks can exploit 'vacant' resource patches that become available in space and over time, and also relocate nutrients gained from such exploitation to other parts of the mycelial network (Boddy 1999). Fungi have other important roles in soil. They act as plant pathogens, they bind soil particles together, thereby enhancing soil structural stability, and they provide a food source to microbial-feeding fauna. Some fungi, the mycorrhizae, form mutualistic associations with plant roots, supplying growth-limiting nutrients such as N and P to the plant, and in return gaining a supply of C and a space free of competition. The three main types of mycorrhizal fungi that infect plants are discussed in Box 2.2. The functional significance of these fungi and of pathogenic fungi for plant communities will be discussed in Chapter 4.

Soil bacteria are conspicuous in that they are extremely abundant and diverse, and they are ubiquitous, inhabiting the most extreme environments and soils. The use of molecular techniques has shown that bacterial diversity in soil is immense, with estimates of diversity being several orders of magnitude greater than previously thought (Torsvik et al. 2002) (Table 2.2). For example, it has been estimated on the basis of analysis of soil DNA that upward of 10,000 genetically distinct prokaryotic types can be present in a handful of soil (Torsvik et al. 2002), and there may be as many as 4 × 10^6

Box 2.2 Mycorrhizal fungi

The term mycorrhiza literally means *fungus-root*, derived from the Greek *mykes and rhiza*, which respectively mean fungus and root. Some 95% of vascular plants belong to genera that are associated with mycorrhizal fungi, which colonize the cortial tissue of roots during active periods of plant growth. The mycorhizal association is characterized by the movement of plant-derived C to the fungus and fungal-acquired nutrients to the plant. Benefits to the plant include enhanced nutrient acquisition, especially of P, and consequently increased growth and improved fitness. Plants that have mycorrhizas also tend to be more competitive and better able to tolerate environmental stress than non-mycorrhizal ones. In return, the fungus gains a habitat that is comparatively free of competition and a supply of photosynthetic C from the plant. Three main types of mycorrhizal association are detailed below, whereas their ecological significance is considered in Chapter 4.

Ectomycorrhizas (EM fungi). These are most abundant in temperate and Boreal forest ecosystems where they infect roots of trees and shrubs, but especially coniferous species that are often exclusively EM. The diagnostic feature of EM associations is the presence of fungal hyphae between root cortical cells to form a complex, intercellular system called the Hartig net. Some fungi also form a sheath or mantle of fungal tissue that completely covers the root. Contiguous with the mantle are hyphal strands that extend into the soil. There are over 4000 species of EM fungi, primarily Basidiomycetes, and more than one species can infect an individual plant.

Arbuscular mycorrhizal fungi (AM fungi). These fungi infect a wide variety of plants, such as grasses, herbs, agricultural crops, and some legumes. The diagnostic feature of AM fungal associations is a highly branched arbuscule within the root cortical cells. The arbuscule is thought to be a site of exchange between the fungus and plant, an assumption that is based on the large surface area of the arbuscule interface. Some AM fungi also have vesicles, which are thin-walled, lipid-filled structures in intercellular spaces that are used for storage. External structures of AM are the hyphae that extend into soil, and spores that function as resting stages and propagules. These are produced asexually from swellings on hyphae in soil. AM fungi are classified in the order Glomales.

Ericoid mycorrhizal fungi. These infect ericaceous plants, for example, heathers and dwarf shrubs, which commonly occur in alpine and tundra ecosystems. These are typically associated with acidic, nutrient poor, highly organic soils. A distinguishing feature of these associations is that the fungus penetrates the root cortex cells, but does not form arbuscules. The fungi also produce a loose welt of hyphae on the root surface, but no mantle is formed. The fungi involved are Ascomycetes.

Table 2.2 Prokaryotic abundance and diversity in different habitats. (Adapted from Torsvik et al. 2002)

Habitat	Abundance (cells cm^{-3})	Diversity (genome equivalents)
Forest soil	4.8×10^9	6000
Pasture soil	1.8×10^7	3500–8800
Arable soil	2.1×10^{10}	140–350

prokaryotic taxa in soil (Curtis et al. 2002). With estimates of as many as 10^{10}–10^{11} bacteria per gram of soil, their abundance is equally astounding (Horner-Devine et al. 2004). Like fungi, heterotrophic bacteria have a general role in soil of degrading a range of organic compounds via the production of extracellular enzymes. Reflecting their taxonomic variety, the metabolic diversity of bacteria is equally immense. This is perhaps best illustrated by the ability of bacteria to degrade many xenobiotic compounds that are toxic to most other organisms in soil. Certain groups of bacteria also perform specific functions of great ecological importance. For example, chemoautotrophic bacteria such as nitrifiers (of the genera *Nitrosomonas* and *Nitrobacter*) are of importance for nutrient cycling in natural and managed systems since they oxidize ammonium to nitrate, the process of nitrification (Chapter 3). Symbiotic nitrogen-fixing bacteria of the genus *Rhizobium* infect roots of legumes, greatly increasing N supply to the system, having important consequences for plant productivity and community structure (Chapter 4). Some free-living, non-symbiotic heterotrophic bacteria also perform this process of N fixation, for example, those of the genera *Azotobacter* and *Clostridium*. As with fungi, the ecological significance of these and other groups of bacteria for ecosystem properties and plant communities will be discussed in later chapters.

2.2.2 Secondary and higher-level consumers

The soil fauna exist alongside microbes, feeding on them and on each other, and on soil organic matter. There is a tremendous variety of animals in soil, with much variation in feeding specialization, life history strategies, and spatial distribution. Species of soil animals also differ markedly in their response to abiotic factors, with reported interspecific variation in their tolerance to soil moisture, food availability, and drought (Scheu and Schultz 1996; Bongers and Bongers 1998), which may determine the species-specific spatial and temporal distribution of biota in soil. Further, many soil animals, like microbes, have the ability to undergo periods of inactivity when conditions are unfavourable, thereby allowing them to tolerate periods of harsh soil conditions. All these factors contribute to making the study of soil animals very challenging, in terms of both their sampling and their characterization into functional groups. As already mentioned, the simplest way of classifying soil animals is on the basis of their body size, which ranges from the smallest protozoa of 1–2 μm to mega earthworms and other macrofauna. This section will focus on some of the important groups of organisms within each of the body size categories, namely microfauna, mesofauna, and macrofauna.

Microfauna

The two most abundant groups of microfauna are the protozoa and nematodes. Soil protozoa are single-celled eukaryotic organisms, which are commonly subdivided into four groups: flagellates, naked amoebae, shelled or testate amoebae (Testacea), and ciliates (Fig. 2.3). As protozoa require a water film for locomotion and feeding, activity is limited to the water-filled pore space in soil, but they can withstand drying of the soil and other adverse conditions by rapidly forming resistant structures called cysts. They can also rapidly emerge from cysts when conditions become favourable, for example, after rains in desert ecosystems (Parker et al. 1984). While protozoan abundance varies greatly in soil, both spatially and temporally, they commonly reach densities of around 10^4–10^6 g^{-1} soil in surface soil, declining to as low as 100 g^{-1} soil in deeper soil horizons where organic matter and prey availability are low (Schnürer et al. 1986). They are also quite diverse: a total of 1092 species of ciliate fauna have been recorded globally (Foissner 1997a), but the maximum number recorded at any one site is 139 species, in the Manaus, Brazil (Foissner 1997b, 1999a).

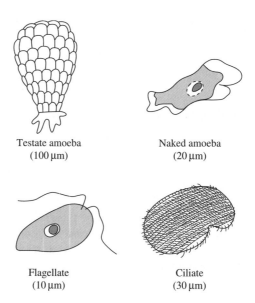

Testate amoeba
(100 μm)

Naked amoeba
(20 μm)

Flagellate
(10 μm)

Ciliate
(30 μm)

Fig. 2.3   Schematic sketch of the four basic groups of protozoa. (Redrawn from Bardgett and Griffith 1997.)

The importance of protozoa in terrestrial ecosystems results mainly from their feeding activities. While the feeding habits of protozoa in soil are still uncertain, most are thought to feed on bacteria, although fungal-feeders, predatory protozoa, and saprophytic protozoa that absorb soluble compounds are also found. The strong link between protozoa and their prey is revealed in studies of their temporal dynamics. For example, amoeba numbers have been shown to increase 5 to 10-fold immediately after rain events because of a stimulation of bacterial prey (Clarholm 1994), and protozoan and bacterial populations have long been known to show classic predator–prey relations when sampled over time (Cutler et al. 1923). These feeding activities have major consequences for ecosystems, affecting nutrient and energy flow both directly and indirectly. Their direct effect stems from the fact that a large proportion (~60%) of ingested nutrients are excreted into the soil environment, enhancing the availability of nutrients for plant uptake, whereas their indirect effect relates to their impact on the activity and composition of microbial communities (Bardgett and Griffiths 1997).

Like protozoa, soil nematodes (roundworms) (Fig. 2.4) require a water film around soil particles in which to move, feed, and reproduce; they also have similarly evolved mechanisms to survive unfavourable conditions (cryptobiosis, a complete cessation of metabolism, and dormant juvenile stages). Nematodes in soil may be grouped according to the type of food that they consume, based on the morphology of their mouthparts (Fig. 2.5). The most common feeding groups of nematodes are plant-feeders, bacterial-feeders,

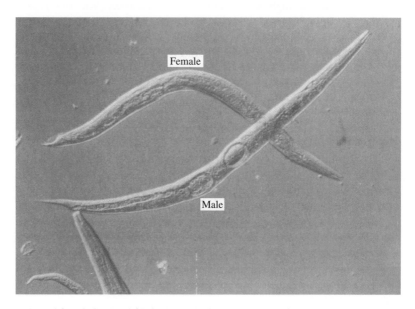

Fig. 2.4   Male and female bacterial-feeding nematodes. (Reproduced from Bardgett and Griffith 1997.)

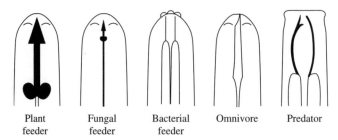

| Plant
feeder | Fungal
feeder | Bacterial
feeder | Omnivore | Predator |

Fig. 2.5   Schematic diagram of mouthparts of nematode feeding groups most commonly found in soil. (Redrawn from Bardgett and Griffith 1997.)

fungal-feeders, omnivores, and predatory nematodes (Yeates et al. 1993). As with protozoa, the role of nematodes in soil systems relates to their feeding activities, and can be direct via the excretion of nutrients into soil or indirect through altering the size, composition, and activity of the microbial community. They also have important roles in the inoculation of new substrates by phoretic transport (i.e. microbes adhering to the surface of larger organisms) or excretion of viable microbes. Plant-parasitic nematodes are increasingly being shown to have important direct effects on plant communities. All these roles will be discussed in more detail in later chapters.

Nematodes are the most numerous multicellular animals in soil, reaching densities as high as 30 million m^{-2} in Oak forest (Volz 1951) and 10 million m^{-2} in highly productive lowland pasture (Yeates et al. 1997a). Even in soils that are polluted with heavy metals, densities can reach 260,000 m^{-2} (Bardgett et al. 1994). Substantially more is known about the diversity of nematodes than that of protozoa. Boag and Yeates (1998), for example, reviewed global data on soil nematode diversity and found that, at 61.7 species per sample, temperate broadleaf forest had the greatest species richness. More recently, Bloemers et al. (1997) recorded a total of 431 species from 24 sites in a Cameroon forest, which is the greatest diversity of nematodes ever recorded. Many species of nematodes also inhabit soils of less fertile ecosystems: an intensive study of a New Zealand sand dune site yielded 44 nematode species under a single colonizer plant, *Ammophila arenaria* (Yeates 1968), while 27 species were found in un-vegetated sand dunes in Scotland (Wall et al. 2002). Even in soils of Antarctic dry valleys, which are thought to be the coldest and driest places on earth, five endemic species of nematodes exist, representing two functional groups (Virginia and Wall 1999).

Mesofauna

The microarthropods are the best known of the mesofauna. The two most abundant microarthropods in soil are the Collembola, which are small (up to 5 mm long), wingless insects, with six abdominal segments and biting mouth-parts which can be withdrawn into the head, and the Acari (mites), which have sac-like bodies, which may be divided by a furrow into anterior and posterior sections (Fig. 2.6). Mites are classified into seven orders, but most soil-forms belong to the orders Crytostigmata, Mesostigmata, Prostigmata, and Astigmata. Numerically, microarthropods are the most abundant non-aquatic faunal group in the soils of most ecosystems. Densities of soil microarthropods can be as high as 300,000 m^{-2} in permanent grassland with dense root systems and high organic matter content (Bardgett and Griffiths 1997). In temperate grasslands, the biomass of mites and Collembola is often reported to be similar, whereas in tropical grasslands the biomass of mites can be 2–5 times that of Collembola (Petersen and Luxton 1982). Many records of microarthropod diversity exist. For example, Behan-Pelletier (1978) recorded a total of 396 mite species across 35 sites in the North American arctic and subarctic, and the same author found 250 species of mites within 1 m^2 of litter and soil, to just 5 cm depth, in mixed deciduous/coniferous forest in Quebec, Canada (Behan-Pelletier, personal communication). Siepel and Van de Bund (1988) found 108

Fig. 2.6   A selection of soil microarthropods extracted from a temperate grassland soils in the United Kingdom. (a) The mite *Uropoda minima*; (b) The collembolan *Friesea mirabilis*; (c) The mite *Lysigamasus truncus;* and (d) The collembolan *Ceratophysella denticulata*. (Images by Lisa Cole.)

species of microarthropods in 500 cm^2 soil of an unmanaged Dutch grassland, and 58 prostigmatid and 59 oribatid species of mites were found in Danish 'poor' pasture (Weis-Fogh 1948). Lower diversities of Collembola have been reported for other grassland systems: 27 species of Collembola were found in tallgrass prairie in Illinois (Brand and Dunn 1998), whereas only 12 species were found in low-productivity, acid grassland in Scotland (Cole and Bardgett, unpublished data).

Another notable, but less widely studied group of mesofauna are the enchytraeids (Oligochaeta), commonly known as potworms (Fig. 2.7). These animals are considered to be 'keystone species' in acid mor soils, such as those of Boreal forest, tundra, and heathland, where they can reach densities of up to 400,000 m^{-2}, constituting some 75% of the total faunal biomass (Didden 1993). Much lower densities of 4000 m^{-2} are found in agricultural and mull forest soils, where earthworms typically dominate the faunal biomass (Didden 1993). They are less conspicuous because of their small size, smaller than other soil annelids like earthworms, being only 10–50 mm long. Their main food source is fungal mycelium, but they also feed on decaying organic matter and microbes associated with it (Didden 1993). There is little information available on the species composition of enchytraeid communities, but there are typically 5–6 species in soils of Scots pine (*Pinus sylvestris*) forests (Lundkvist 1983; Didden and de Fluiter 1998). Higher levels of enchytraeid diversity have been found; for example: Chalupský (1995) recorded 23 species across a range of Czech mountain forest soils, Standen (1984) found 28 species in hay meadows in northern

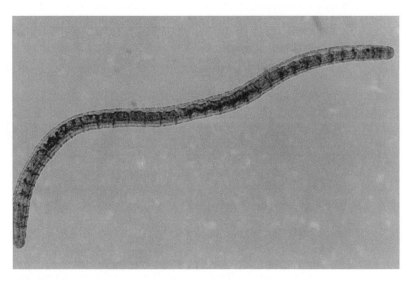

Fig. 2.7   Enchytraeid worm, the primary detritivore of acidic peat soils of Boreal forest, heathland, and northern peatland. (Image by Lisa Cole.)

England, and Ștefan (1977) identified 35 species in a range of sites in the Cerna Valley, Romania.

Macrofauna

The macrofauna are the most conspicuous fauna of soil, and also the most documented in terms of their biology and impact on soil fertility. They include a wide array of taxonomic groups that cross several trophic levels. For example, the millipedes and woodlice are major consumers of organic debris in forests, and various insect larvae, such as those of the crane fly (Tipulidae), are important consumers of root material, whereas centipedes, spiders, scorpions, and beetles tend to be the dominant predators in soil and litter. Ants are another group of soil macrofauna and are among the most ubiquitous insects on the planet. They have a variety of feeding habits, being predators, microbial- and plant-feeders, and opportunistic omnivores. In the biomes where ants are abundant, they affect many soil processes through their nesting activities. These nests can be >2 m in height, at densities between 200 and 1000 mounds ha^{-1}, and they contribute greatly to the creation of the patch mosaics that characterize the soils and vegetation of many landscapes. These nests also greatly affect critical ecosystem processes such as nutrient cycling and water redistribution.

Earthworms are probably the best known of the macrofauna and there is a huge literature on their biology and importance for soil fertility in natural and agricultural systems (Lee 1985; Edwards and Bohlen 1996; Edwards 2004). Indeed, literature on their beneficial roles for soil fertility dates back as far as Aristotle, who called them the 'intestines of the earth'; Charles Darwin spent many years observing their major influence on the physical nature of soil (Darwin 1881). Earthworms usually make up most of the animal biomass in soil, especially in productive ecosystems with mull soils such as temperate grasslands, deciduous forests, and tropical pastures and rainforests, where densities can reach 400–500 m^{-2}. However, in acid mor soils typical of Boreal forest and heathland, earthworms are virtually absent, and are typically replaced by enchytraeid worms as the dominant taxon. Also, earthworms generally do not occur in deserts and arid grasslands where water is limited. A review of earthworm species present under various vegetation types around the world by Lee (1985) showed the typical diversity to be only 2–5 species, with a maximum of 11. However, a worldwide survey of earthworms in tropical regions reported that a total of 51 exotic and 151 native species commonly occur in tropical agro-ecosystems (Fragoso et al. 1999b).

Another group of macrofauna considered as 'keystone species' in the tropics and subtropics are the termites. As with earthworms, much has been written about their biology and important roles in these ecosystems (see Aber et al. 2000). Termites feed mainly on decaying organic matter and

(a)

(b)

Fig. 2.8   Termite mounds:(a) Mulga country near Alice Springs, Australia, where the landscape is dotted with dead wood (the major resource) and the conical mounds of *Amitermes vitiosus*. (Reproduced with permission from Marcel Dekker; Image by David Bignell); and (b) mounds of the soil-feeding termite *Cubitermes severus* in the moist savanna landscape of Nigeria. (Reproduced with permission from Kluwer Academic Press; Image by Reine Leuthold.)

roots, and are best known for their dramatic effects on soil nutrient cycling and the physical architecture and hydrology of soils, through their mound building and gallery excavating activities (Fig. 2.8). Most termites depend on symbiotic gut bacteria for energy metabolism, whereas others, the so-called 'fungus growers', lack these gut flora and rely on externally cultivated basidiomycete fungi for their nutrition. Some termites also contain in their guts populations of flagellate protozoa that have the ability to degrade cellulose and other plant polysaccharides. Others have the capacity to fix atmospheric N through the action of their gut bacteria, and as a result they are often the primary providers of N in many soils and food chains, with termite mounds serving as foci for nutrient redistribution in some landscapes. Termite abundance varies from <50 m^{-2} in arid savannas to more than 7000 m^{-2} in some African forests, where they make up some 95% of the insect biomass in soil (Aber et al. 2000). More than 2300 species have been described in 270 genera and 6 families, but spatial patterns of termite diversity are strongly influenced by factors such as latitude, being maximal at the equator, and rainfall, with communities being more species rich and abundant at the same latitude where rainfall is greater (Aber et al. 2000).

2.3   Patterns of soil biodiversity

It is fascinating to note that biological diversity is not the same everywhere. There is almost infinite variation in patterns of diversity across and within the Earth's ecosystems. Furthermore, these patterns change constantly with time; the diversity assessed at a particular locality will vary greatly both within and between years, but will also change over longer timescales as a result of processes of succession and evolutionary change. This tremendous spatial and temporal complexity in diversity, which occurs both above-ground and below-ground, makes it extremely difficult, if not impossible, to evaluate and explain patterns of species richness. However, carefully framed questions about species diversity, which recognize the existence of such complexity, can give insights into the wide range of processes that control diversity on a range of spatial and temporal scales. This section examines some of the known features of diversity in soils, at global and local scales, and discusses how such diversity might change in particular locations over time.

2.3.1 Global patterns of soil biodiversity

A host of global patterns of spatial variation in biodiversity have been explored, but perhaps the most prominent and intensively studied are lati-tudinal gradients of species richness. It has long been recognized that the number of species in most taxonomic groups is lowest at the poles and increases towards the tropics (Huston 1994). However, information on latitudinal gradients of taxonomic groups is almost entirely based on above-ground organisms, with virtually nothing being known about latitu-dinal patterns of species that live in soil. As far as the author is aware, no studies have comprehensively assessed the diversity of soil biota across lat-itudinal gradients using either standardized sampling approaches or simi-lar taxonomic resolution. It is therefore rather dangerous to speculate too much about latitudinal patterns of soil biodiversity. These problems are clearly shown by the study of Boag and Yeates (1998). The authors synthe-sized information on the species richness of soil-inhabiting nematodes and showed that nematode diversity was substantially lower near the poles than in tropical or temperate forests (Fig. 2.9). Their data also showed that nematode diversity was greater in temperate than tropical forests, suggest-ing a non-linear relationship between nematode diversity and latitude, which contrasts with what occurs above-ground. The problem with this relationship, however, is the gross imbalance in soil sampling in favour of temperate latitudes (80% of samples used were taken from between 40 and 60°N), with very few studies having been made in the tropics or polar regions. Meaningful comparisons of nematode diversity between regions of the world are therefore not really possible, other than towards the poles, where the diversity of nematodes (Boag and Yeates 1998; Wall and Virginia 1999), protozoa (Smith 1996), and earthworms (Lavelle et al. 1995) (Fig. 2.10(a)) drops substantially, presumably as a result of reduced resource availability and harsh environmental conditions that require special adaptations, especially cold and desiccation tolerance (Bale et al. 1997). The only group of soil animals that does appear to follow a clear latitudinal gradient is the termite group, which shows an increase in diver-sity towards the tropics (Eggleton and Bignell 1995) (Fig. 2.10(b)).

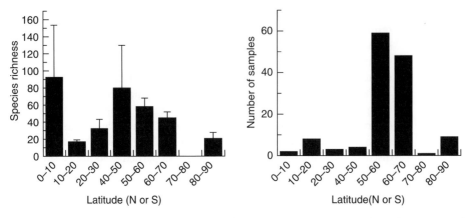

Fig. 2.9   Latitudinal patterns of soil nematodes and sampling distribution. (Data from Boag and Yeates 1998.)

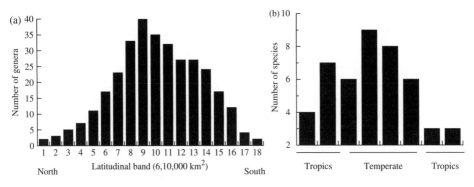

Fig. 2.10   Latitudinal patterns of (a) termites (data from Eggleton et al. 1994), and (b) earthworms (data from Lavelle et al. 1995).

Some ecologists actually doubt whether latitudinal patterns of soil biological diversity exist at all, at least across temperate and tropical regions. The rationale for this is that many soil organisms, especially the microfauna and microbes, are cosmopolitan, being able to migrate extensively across the globe (Roberts and Cohen 1995; Finlay et al. 1999). Furthermore, the abundance of individuals of free-living microbial species is so large that their dispersal is rarely (if ever) restricted by geographical barriers (Finlay 2002). That soil biota have such a 'ubiquitous' nature is evidenced by a recent study which revealed that approximately one-third of the global diversity of soil protozoa was present within a single patch of grassland in Scotland, and that those species that were locally rare were similarly rare on a global scale (Fig. 2.11) (Finlay et al. 2001). That certain taxonomic groups of soil biota are cosmopolitan, however, has been contested because some

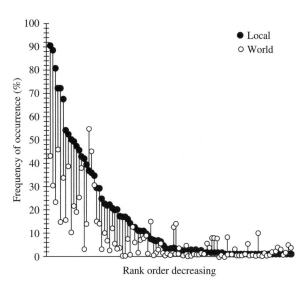

Fig. 2.11   Frequency of detection of 95 ciliate species in 150 soil samples from 1 ha grassland in Scotland, and of the same species in 606 samples taken worldwide, showing a significant agreement between local and global diversity, that is, that those species that were locally rare were similarly rare on a global scale. (Redrawn with permission from Finaly 2002, Science 296, 1061–1063. Copyright 2002, AAAS)

free-living ciliates (Foissner 1999*b*) and heterotrophic soil bacteria (Cho and Tiedje 2000) are known to have restricted biogeographical distributions. Also, estimates of global soil biodiversity are considered to be highly speculative because only a tiny fraction of potential habitats across the globe has been carefully analysed (Foissner 1999*b*).

2.3.2   Landscape patterns of soil biodiversity

At the landscape level, patterning of soil biota is mostly related to factors such as soil pattern and topography (Chapter 1), and to factors such as disturbances and vegetation patchiness. At the broadest level, it has been argued by Ponge (2003) that the characterization of soil into mull and mor humus types provides a framework for understanding the distribution of soil biodiversity. Fertile mull soils associated with productive ecosystems, such as deciduous forest and temperate grassland, tend to support a high level of plant, animal, and microbial diversity owing to ample provision of resources and a high level of heterogeneity, caused largely by the activities of the organisms themselves. In contrast, the harsh climatic conditions and abundance of recalcitrant organic matter typical of unproductive, mor type soils mean that fewer species are present (Ponge 2003). This kind of patterning is also evident on a more local scale, such as in forests, where differences in the quality of litter beneath coexisting tree species produce zones of influence that explain the patchy distribution of soil organisms and process rates (Saetre and Bääth 2000). Similarly, in semiarid ecosystems it has been shown that patterns of soil biodiversity are strongly related to the patchy spatial pattern of vegetation. At these sites, biodiversity in soil below plants is greater than in adjacent exposed soil, suggesting that plants serve as 'resource islands' providing resources for soil biota (Herman et al. 1995). In contrast, in ecosystems where vegetation is sparse, abiotic factors determine spatial patterns of soil biodiversity. For example, in the McMurdo Dry Valleys site in Antarctica, patterns in the diversity of soil nematode communities across the landscape depend on abiotic factors such as organic C, salts, and soil moisture (Courtright et al. 2001) (Fig. 2.12).

Another important factor affecting the distribution of soil biodiversity at the landscape level is physical disturbance, which generally leads to dramatic reductions in soil biodiversity. There are numerous such cases. For example, disturbance of soil through tillage has been shown to reduce the abundance and diversity of earthworms (Springett 1992), although the

Fig. 2.12   Dry Valleys in Antarctica. Here, patterns in the diversity of soil nematode communities across the landscape depend on abiotic factors such as organic C, salts, and soil moisture. (Image by David Hopkins.)

scale of these effects depends on soil type, climate, and tillage operation (Chan 2001). Other studies have examined gradients of disturbance resulting from the conversion of natural vegetation to agriculture, showing that the diversity of certain soil biota is reduced as a consequence. For example, the conversion of primary tropical forest to agriculture has been shown to dramatically reduce microbial biomass (Dlamini and Haynes 2004), and the diversity of macrofauna (Lavelle and Pashanasi 1989), termites (Eggleton et al. 2002), microarthropods (Tain et al. 1992), and nematodes (Bloemers et al. 1997) (Fig. 2.13). Further, studies done in many parts of the world report that earthworm communities of agro-ecosystems have lower species richness and a lower number of native species than do undisturbed ecosystems (Fragoso et al. 1997; Dlamini and Haynes 2004). The diversity of Collembola is also commonly found to be greater in native prairie than in prairie that has been influenced by agriculture (Brand and Dunn 1998), and the diversity of soil nematodes has been shown to decrease with agricultural improvement of native pastures in New England Tablelands (Yeates and King 1997). Measurement of changes in the relative abundance, or evenness, of phospholipid fatty acids (PLFA) suggests that the broad scale phenotypic diversity of microbial communities in soil is also reduced as a consequence of disturbances caused by intensive agricultural management of temperate grasslands (Bardgett et al. 2001*b*) (Fig. 2.13). Studies of mycorrhizal fungi also reveal that disturbances lead to dramatic declines in species richness. For example, Daniell et al. (2001) used molecular tools to show that AM fungal diversity is much lower in arable systems than in adjacent, undisturbed woodlands. Similarly, species diversity of EM fungi is reduced by clear-cutting of forests (Hagerman et al. 2001) and late successional forests have more diverse mycorrhizal communities than do earlier successional forests that have been subjected to natural or anthropogenic disturbance (Horton and Bruns 2001). The depletion of species richness resulting from conversion of land to farming is also reinforced by comparison of the number of AM species typically found in arable lands (6 species; Helgason et al. 1998) with the number found in semi-natural grasslands (24; Vanderkoornhuyse et al. 2002) and tropical forest (30; Husband et al. 2002).

It is important to note that there are likely to be numerous, interacting reasons for declines in diversity of soil biota resulting from the conversion of natural ecosystems for agriculture. Physical disturbance resulting from site preparation (e.g. deforestation, burning, and cultivation) and declines in the amount and complexity of organic residues returned to the soil are likely to be the most important factors (Beare et al. 1997*b*). However, conversion of native vegetation to agricultural land also typically results in a substantial loss of soil organic matter owing to the small amounts of organic matter returned to soil, and regular tillage favouring enhanced organic matter mineralization. This is especially the case in farming systems

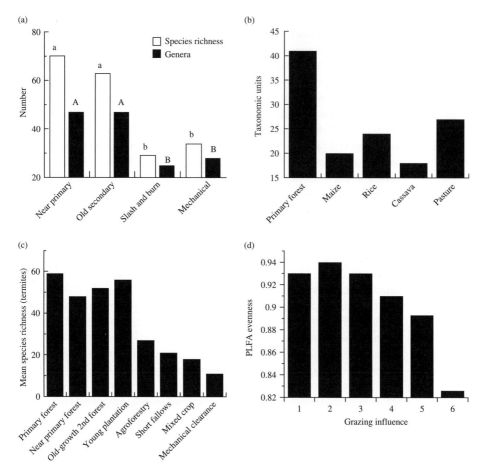

Fig. 2.13 Effects of anthropogenic disturbances on the diversity of soil biota: (a) effects of forest disturbances on the number of species and genera of nematodes in soil of tropical forests, Cameroon (data from Bloemers et al. 1997.); (b) effects of land management on the number of taxonomic units of macrofauna in Peruvian Amazonia (data from Lavelle and Pashanasi 1989); (c) termite diversity across a gradient of anthropogenic disturbance in the humid forest zone of West Africa (data from Eggleton et al. 2002); and (d) changes in PLFA evenness index of soil microbial communities along a grazing-related disturbance gradient on grasslands in the Snowdonia National Park, UK. Site 1 = ungrazed control, 2 = long-term ungrazed, 3 = short-term ungrazed, 4 = lightly grazed, 5 = moderately grazed, and 6 = heavily grazed (data from Bardgett et al. 2001b).

where crop residues are burned rather than left as mulch (Dominy et al. 2002; Graham et al. 2002; Dlamini and Haynes 2004). The organic matter returned to soil in agricultural systems is also of much lower complexity than that in native ecosystems, often being from a single crop; as will be discussed in Chapter 4, the diversity of decomposers is typically greater in more complex litter mixtures than in simple soil substrates (e.g. Hansen and Coleman 1998; Hansen 2000). This is also shown by the observation that declines in termite diversity with increasing intensification from primary forest to farmland are strongly linked to diminished plant community complexity (Gillison et al. 2003; Jones et al. 2003) (Fig. 2.14), which is a strong determinant of the spatial and temporal heterogeneity of plant detritus (Beare et al. 1995). Declines in decomposer diversity resulting from the conversion of natural ecosystems to farmland may also be linked to changes in soil microclimate. For example, Critchley et al. (1979) studied relationships between soil microclimate and soil invertebrate populations in cultivated and native bush plots in the humid tropics, and found that diurnal temperatures in cultivated soils ranged from 26 to 32 °C, whereas in the bush soils the temperature was almost constant at 25 °C. Bush soils also had a higher moisture content than cultivated ones; consequently, the activity and abundance of most surface-dwelling fauna was greater in native bush than in adjacent cultivated areas (Table 2.3).

Disturbance does not always have negative effects on soil biota. Indeed, some disturbed soils can have highly diverse soil communities, especially when the disturbances enhance spatial heterogeneity, thereby providing opportunities for more species to coexist. A good example of this is soils of the riparian zone (the zone between terrestrial and aquatic ecosystems) that possess an unusually high diversity of fauna and microflora (Ettema et al. 1999, 2000). This diversity is maintained by a variety of disturbances (e.g. periodic flooding, drought, freezing, abrasion, erosion, and occasionally toxic concentrations of nutrients) that create a spatial and temporal mosaic with few parallels in other systems (Ettema et al. 2000).

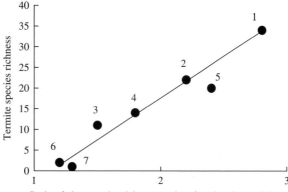

Fig. 2.14 Relationships between termite species richness across a forest disturbance gradient in lowland Sumatra and the ratio of plant species richness to plant functional types. The numbers next to data points indicate the land use systems:1. Primary rain forest; 2. Logged over rain forest; 3. Industrial softwood plantation; 4. Eight-year-old rubber plantation; 5. Old growth jungle rubber tree mosaic; 6. *Imperata* grassland; 7. 10-year-old cassava plantation (data from Gillison et al. 2003 and Jones et al. 2003, derived from Giller et al. 2005).

Table 2.3 Effects of cultivation of native bush plots on populations of microarthropods in a tropical Alfisol (0–10 cm). (Adapted from Critchley et al. 1979)

Land use	Native bush plots	Cultivated plots
Acari		
Cryptostigmata	25,098	4221
Prostigmata	14,830	17,701
Mesostigmata	4570	1887
Astigmata	1609	725
Collembola		
Isotomidae	10,607	2158
Entomobryidae	2826	1179
Onychiuridae	1982	348
Sminthuridae	1408	942
Poduridae	865	63

A note on above-ground theory

One intriguing question concerning landscape patterns of diversity is whether the same determinants of above-ground diversity operate below-ground. There are two factors that have traditionally been emphasized as key determinants of above-ground diversity: productivity or resource supply and consumption or physical disturbance (Fig. 2.15). Species richness is often found to be unimodally related to productivity, such that peak diversity is at intermediate productivity: the humpbacked model (e.g. Grime 1973; Al Mufti et al. 1977; Grace 1999), with declining diversity at higher levels of productivity being due to competitive exclusion. Competitive exclusion can, however, be prevented by periodic mortality events that are caused by consumption or physical disturbance (Connell 1978; Huston 1979); these factors, likewise, display unimodal relationships with diversity, as hypothesized by the 'intermediate disturbance hypothesis' of Connell (1978). There are insufficient data available to test either of these ideas with any rigour in soil, and those available do not appear to lend much support for either the productivity–diversity or the disturbance–diversity relationship operating in soil. It is known that productivity is an important determinant of soil food-web structure at the extremely unproductive end of productivity gradients, for example, in caves, where detrital food chains become longer and more diverse in productive situations, with more energy being diverted to higher trophic levels (Moore and de Ruiter 2000). However, in a literature synthesis Wardle (2002) found no data to support the idea that diversity declines along the most favourable portion of the productivity gradient, that is, that soil organism diversity peaks at intermediate levels of productivity, thereby failing to conform to the humpbacked model. If the humpbacked model does not apply for soil

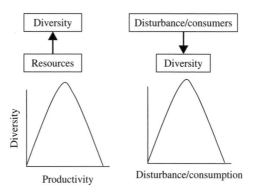

Fig. 2.15   Disturbance/consumer versus resource control of species diversity. Both models predict that diversity will be maximal at intermediate levels of disturbance and productivity.

biota it suggests that below-ground communities could differ from their above-ground counterparts, in that soil biodiversity is not so strongly regulated by competition, and competitive exclusion does not occur when resource availability in soil is increased (Wardle 2002). Similarly, there is little support in the literature for optimization of soil animal diversity at intermediate levels of disturbance (see Bardgett et al. 2005); on the contrary, as already discussed, disturbances generally lead to dramatic reductions in soil biodiversity.

Another important determinant of above-ground diversity that has long been recognized is island size, which is part of the island biogeography theory of McArthur and Wilson (1967). Island size determines, among others things, the amount of resources available for organisms, and in general, larger islands have a greater variety of resources and hence more species— the so-called species–area relationship. Wardle et al. (2003a) tested these ideas for decomposer communities by examining the diversity of groups of biota in the organic matter of 'suspended soils' produced by spatially separated epiphytes or treetop 'islands' in the crowns of canopy tree species in old-growth, warm, temperate forest in Northland, New Zealand. They found that larger 'islands' supported a greater diversity of macrofauna and microarthropods, which is consistent with the island size theory. It was also found that physical proximity of islands to sources of colonizing organisms did not influence the diversity of these faunal groups, which contrasts with the predictions of the island biogeography theory (Wardle et al. 2003a).

2.3.3   Local patterns of soil biodiversity

Local-scale patterns of soil biodiversity, over ranges of centimetres to metres, are most likely to be related to habitat heterogeneity, in terms of both the structural complexity, or patchiness, of the soil environment and the chemical complexity of resources. As discussed in Chapter 1, the soil is extremely variable in its structure and composition, and all soil systems are spatially heterogeneous across many orders of magnitude of scale. This provides unrivalled opportunities for organisms to establish niches and coexist (Anderson 1975; Ettema and Wardle 2002; Bardgett et al. 2005), but also leads to extreme spatial variation in the community-level biological properties of soil (Ritz et al. 2004) (Fig. 2.16). This notion is supported by a number of studies that show that local spatial heterogeneity promotes soil biodiversity and that soil biological communities display strong niche differentiation, thereby enabling species to coexist. For example, on the issue of the role of spatial heterogeneity, Zhou et al. (2002) showed that bacterial diversity was lower in soils saturated with water than in unsaturated soils that were patchier. Also, mixtures of litters have been shown to

(a)   (b)   (c)   (d)

(e)

Fig 2.16   The subterranean habitat: soil structure across five orders of magnitude. (a) field scale; (b) profile scale (bar = 1 cm); (c) root scale [bar = 3 mm]; (d) pore scale (false-colour image of soil thin section, bar = 1 mm); (e) microscale (computer-aided tomographic slices, bar = 500 μm). Note that the physical size of the soil biota spans five similar orders of magnitude. (Image by Karl Ritz, from Ritz 2005).

support more diverse mite communities than single-species litters, owing to an increased range of food resources and increased habitat complexity (Gill 1969; Stanton 1979; Hansen and Coleman 1998; Hansen 2000), and impacts of litter mixing on diversity have been shown to disappear as litters decompose and resources become more morphologically homogeneous (Hansen 2000). Mite diversity has also been found to be positively related to microhabitat complexity of woodland soil and litter layers (Anderson 1978), and species diversity of litter-dwelling gastropods has been shown to increase with increasing plant diversity (Barker and Mayhill 1999).

On the issue of niche differentiation, bacteria and fungi are known to have numerous adaptive strategies in terms of spore types, physiological modes of adaptation to resource limitation, sexual and asexual reproductive modes, and eco-physiological tolerance. Similarly, soil animals show tremendous specialization in terms of their feeding and life history strategies (Siepel 1996; Scheu and Falca 2000) and their spatial distribution, enabling them to avoid competition. For example, by studying the natural variation in stable isotope ratios ($^{15}N/^{14}N$) (Box 2.3) of mite species extracted from forest soils, Schneider et al. (2004) revealed strong trophic niche differentiation in microcroarthropod communities, with orabatid mites spanning three to four trophic groups. Soil animal species also differ markedly in their response to abiotic factors, with much interspecies variation in their tolerance to soil moisture, food availability, and drought (Verhoef and van Selm 1983; Scheu and Schultz 1996; Bongers and Bongers 1998), which may determine the species-specific spatial and temporal distribution of biota in soil.

Roots also create heterogeneity in soil, through effects on soil physical structure, and also via their death and decomposition, and through the exudation of easily degraded compounds into the zone of soil immediately adjacent to the root, the rhizosphere. Those compounds lost from plant roots by exudation, which include low molecular mass exudates and secretions, polysaccharide mucilages, and lysate, represent a high-quality nutrient source for the growth and maintenance of rhizosphere microorganisms (Whipps and Lynch 1983). This leads to enhanced microbial growth in the rhizosphere relative to bulk soil, and in turn increases prey availability for microbial-feeding fauna, such as protozoa and nematodes, which are often at their maximum abundance in the root zone (Bardgett and Griffiths 1997). A large proportion of photoassimilate C is allocated to soil via root exudation (5–10%) (Farrar et al. 2003) (Fig. 2.17), and of that a significant quantity can also be allocated to mycorrhizal fungi, especially when soils are P limited (Johnson et al. 2003c). Many factors such as plant identity, physiological state, age, and environmental conditions, alter the quantity and chemical make-up of C compounds released from roots into the rhizosphere (Bardgett et al. 1998), which in turn leads to differential effects on the soil microbial community and food-web structure. The significance of

Box 2.3   Using stable isotopes to estimate trophic position of soil animals

There is presently much interest in using stable isotopes, particularly of C and N, to study the structure and dynamics of ecological communities (Post 2002). Stable isotope techniques can provide a continuous measure of trophic position that integrates the assimilation of energy or mass flow through different trophic pathways that lead to an organism. In particular, the ratio of stable isotopes of $N(\delta^{15}N)$ can be used to estimate the trophic position of an organism since the $\delta^{15}N$ of consumer organisms is typically enriched by a mean value of 3.4 δ units relative to its diet. Historically, this trophic fractionation of 3.4‰ was thought to originate from the excretion of isotopically light N, but the consensus now is that it most likely originates from a combination of isotopic fractionation during both assimilation and protein synthesis, and during the excretion of N (Post 2002). While a mean trophic fractionation of 3.4‰ is widely used, it is important to note that it is valid only when applied to entire food web, with multiple trophic pathways. Studies that attempt to quantify trophic structure across just a few feeding links should be cautious, since any single trophic transfer can range between 2% and 5% (Post 2002).

A number of authors have used this approach to study the trophic structure of soil animal communities. For example, Scheu and Falca (2000) studied $^{15}N/^{14}N$ ratios of macro- and meso-fauna of two beech forests, and found that ratios varied strongly, spanning some 15 $\delta^{15}N$ units. Since trophic levels are assumed to differ by 3.4 $\delta^{15}N$ units, this was taken to indicate that the soil food web of these forests spans more than four trophic levels. However, average $\delta^{15}N$ values of the majority of species varied little, forming a cluster of saprophagous or microphytagous taxa that formed a continuum from primary to secondary consumers. The other cluster of organisms was predatory taxa, whose $\delta^{15}N$ value exceeded that of the above by 3.9 and 4.4‰ in the two forests. In view of this, it was concluded that stable isotope analysis may only be useful for depicting very general trophic groupings, such as predators and detritivores. Ponsard and Arditi (2000) came to similar conclusions. They used stable isotope approaches to study food-web structure of soil macrofauna of deciduous forests in France and found that the soil community could only be subdivided into two trophic levels: detrivores that feed on superficial litter, and predators that feed on the detritivores. In view of this, it was concluded that stable istotopes are unable to provide a detailed picture of the trophic links between individual species in soil food webs. This view, however, has recently been challenged by

Schneider et al. (2004). These authors investigated $^{15}N/^{14}N$ ratios of 36 orabatid mite taxa from a range of beech forests in Germany and found that the ratio spanned 12 $\delta^{15}N$ units. This was taken to suggest that different species occupy different trophic niches and that orabatid mites span three or four trophic levels.

While the use of stable isotopes clearly has much potential for offering insights into food-web structure of soils, they must be used with care for a number of reasons. First, experimental information on the degree of enrichment per trophic level is scarce, and for soil animals this information is virtually absent. Second, enrichment is known to vary among species and differ between animals that are starving or non-starving (Schmidt et al. 1999*b*). Third, stable isotope ratios of species may vary in time due to changes in food resources, and also in space, for example, with soil depth. It is well known that $^{15}N/^{14}N$ ratio of soil organic matter increases with soil depth (Nadelhoffer and Fry 1994), and this may affect the ratio of soil organisms colonizing deeper soil layers.

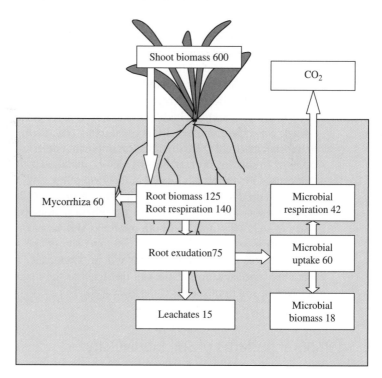

Fig. 2.17   The partitioning of C within the plant and the distribution of root exudates within the soil after net fixation of 1000 units of C. (Redrawn with permission from the Ecological Society of America; Farrar et al. 2003.)

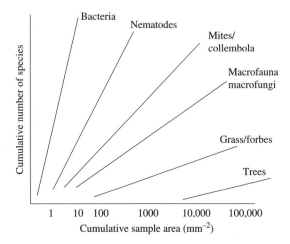

Fig. 2.18   Hypothetical species/sample size relationships for soil organisms and vegetation cover affecting soil habitats. (Adapted from Bardgett et al. 2005)

these factors and also the issue of how variations in plant diversity and composition affect soil biota will be discussed in detail in Chapter 4.

It is important to note that the complexity of the soil environment occurs at a range of spatial scales. Therefore, heterogeneity will potentially influence the diversity of different groups of organisms in different ways, depending on their body size and the size of their habitat unit, or domain. (Fig. 2.18). Microscale heterogeneity, for example, at the scale of the soil particle can be extremely high, providing steep environmental gradients in oxygen concentration over a few millimetres (Fig. 2.16). Such microscale heterogeneity could therefore act as an important determinant of microbial diversity in soil. In contrast, heterogeneity over larger spatial scales, such as that caused by resource 'hotspots' in the form of buried litter, dead animals, roots, and other organic materials, or patterns of root C flow, and also variations in soil structure that affect the physical dimensions of habitable space in soil, will impact on both microbes and larger organisms. In sum, variations in spatial and temporal heterogeneity of the soil environment, from the soil particle scale to the soil horizon scale, coupled with the incredible specialization of the soil biota, act as the major determinant of patterns of soil biodiversity at the local scale.

2.4   Temporal patterns of soil biodiversity

Soil communities are not static. They change constantly over time, ranging from daily to seasonal changes in the abundance of various groups of organisms, and to changes in soil biodiversity over successional time.

Temporal patterns of soil biodiversity are also complicated by the fact that many soil organisms can undergo long periods of inactivity when conditions are unfavourable, thereby allowing them to tolerate periods of harsh soil conditions. The degree to which soil communities change over time is also highly context dependent, varying due to a range of factors, such as the nature and frequency of disturbance regimes, vegetation change, and changes in soil and climatic conditions. This is perhaps most evident in managed systems, where changes in management can affect soil biota over different timescales. For example, effects of changing management regimes (e.g. the cessation of fertilizer application) on soil microbial communities of temperate grasslands in northern England, being linked to vegetation change (Smith et al. 2003), took nine years to detect. Whereas in more nutrient poor grassland, impacts of similar management treatments on microbial communities became evident within only four years, due to rapid changes in soil physico-chemical conditions (Bardgett et al. 1996). Yearly variation in soil biotic communities are also highly evident, for example, over a 23-year period Yeates et al. (2000) observed unexplained, apparently chaotic, changes in nematode assemblages in pasture soils. Over a seven-year period, there was also much variation in soil-associated arthropods under various weed management treatments for *Zea mays* and *Asparagus officinale* crops, which appeared to be related to variations in weed biomass, which itself varied from year to year (Wardle et al. 1999*b*). These studies all emphasize the need for careful thought about how soil biodiversity is evaluated in field experiments and at specific locations.

Soil biodiversity also changes over timescales of tens or hundreds of years, through processes of succession. A number of studies give insights into the kinds of patterns that occur in soil communities over these long timescales. For example, for soil microbes it is evident that during the initial stages of succession—the build-up phase—and towards maximal plant biomass (i.e. community climax), soil microbial communities become increasingly abundant and active (Insam and Haselwandter 1989; Ohtonen et al. 1999), and more diverse (Schipper et al. 2001) and fungal dominated (over bacteria) in nature (Frankland 1998; Ohtonen et al. 1999; Bardgett 2000) (Fig. 2.19). There are also changes in mycorrhizal communities as succession proceeds. In early succession, ruderal plants are generally non-mycorrhizal, whereas in mid-succession the dominant herbaceous plants tend to have a facultative requirement for arbuscular mycorrhizal fungi, and in climax communities the trees and shrubs, which dominate the vegetation, often have an obligate need for ectomycorrhizae (Read 1994). Similar shifts in microbial community composition also appear to occur during secondary succession, for example, after abandonment of management on agricultural grasslands that typically leads to a shift in the composition of the microbial community towards fungal dominance over bacteria (Bardgett and McAlister 1999). It is not entirely clear why these patterns in microbial communities occur, but

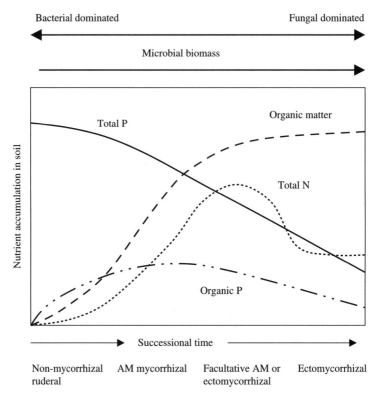

Fig. 2.19   Schematic diagram of changes in below-ground communities during primary succession.

they are most likely to be related to a build-up in the amount and complexity of organic matter, and changes in the quality of resource inputs to soil resulting from vegetation change. They may also be related to changes in the physico-chemical nature of soils, for example, a drop in soil pH that commonly occurs in late succession (Bardgett et al. 2001*b*).

Little is known about how the faunal community changes during succession, but one might expect that fungal-feeding fauna become increasingly dominant, and that predators become more abundant as more energy becomes available for higher trophic levels. However, a study of different stages of secondary succession (from a wheat field to a beechwood) by Scheu and Schulz (1996) showed that soil invertebrates of similar trophic groups responded very differently to successional changes, suggesting that responses of higher trophic levels to succession are difficult to predict. A study by Kaufmann (2001) of surface-active micro- and macroarthropods in the Central Alpine glacier foreland of the Rotmoostal (Obergurgl, Tyrol, Austria) also indicates that in this group patterns of invertebrate succession are not as might be predicted: although faunal colonization and succession in the Alpine glacier

Fig. 2.20   The glacier foreland of the Rotmoostal (Obergurgl, Tyrol, Austria) where initial colonizers are almost exclusively predators, and herbivores and decomposers appear later.
(Image by Otto Ehrmann.)

foreland generally followed a predictable pattern, the first colonizers were
almost exclusively predators, and herbivores and decomposers appeared
later (Fig. 2.20). Similarly, spiders are the earliest colonizers of newly
exposed moraine substrates on the glacier foreland of the Midre Lovénbreen
at Ny-Ålesund, Spitsbergen, Svalbard (78°N), where their densities are highly
correlated with allochthonous inputs of potential prey items, predominantly
chironomid midges (Hodkinson et al. 2001). In these systems, large
allochthonous inputs of insects potentially provide significant quantities of
N and P to the developing ecosystem from the earliest stages of succession,
even before a conspicuous cyanobacterial crust has formed. That spiders
entrap nutrients in such a way could be of high importance in early ecosystem development in these extreme environments.

2.5   Conclusion

There is a vast diversity of organisms that live in the soil and understanding
the factors that regulate this diversity and its consequences for ecosystem

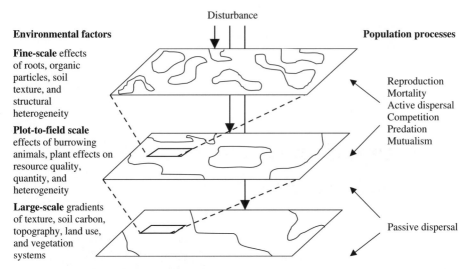

Disturbance

Environmental factors

Fine-scale effects
of roots, organic
particles, soil
texture, and
structural
heterogeneity

Plot-to-field scale
effects of burrowing
animals, plant effects on
resource quality,
quantity, and
heterogeneity

Large-scale gradients
of texture, soil carbon,
topography, land use,
and vegetation
systems

Population processes

Reproduction
Mortality
Active dispersal
Competition
Predation
Mutualism

Passive dispersal

Fig. 2.21   Determinants of spatial heterogeneity of soil organisms. The figure shows that spatial heterogeneity in soil organisms is distributed on nested scales, and is shaped by a spatial hierarchy of environmental and biological factors, and disturbance. (Redrawn with permission from Elsevier Ltd. Ettema and Wardle 2002.)

function, represents a key challenge for soil ecology. The main problem that soil ecologists face in trying to understand and develop theory on patterns and determinants of soil biodiversity is the dearth of information that is available on the diversity of soil biota, especially at the species level, across different spatial and temporal scales (Fig. 2.21). Despite this, we can make certain judgements. In particular, it would appear that patterns and determinants of below-ground diversity differ from those that operate above-ground, especially in relation to the importance of competition and the prevention of competitive exclusion that are important drivers of vegetation change. In soil, patterning of biodiversity is related primarily to the heterogeneous nature, or patchiness, of the soil environment that occurs at different spatial and temporal scales. This heterogeneity provides unrivalled potential for niche partitioning, or resource and habitat specialization, leading to avoidance of competition and hence coexistence of species. The challenge for soil ecologists, therefore, is to identify the hierarchy of controls on soil biological diversity that operate at different spatial and temporal scales, and to determine the role of spatio-temporal patterning of soil biodiversity as a driver of above-ground community assembly and productivity.

3 Organism interactions and soil processes

3.1 Introduction

In the previous chapter, the main players in the soil food web and the factors that regulate patterns of soil biodiversity, were introduced. The next issue to be considered is the functional significance of these organisms and their interactions with one another for ecosystem processes, especially nutrient cycling. Decomposer organisms are essential for the functioning of terrestrial ecosystems largely because they decompose dead organic material in soil, converting this into carbon dioxide and other soluble nutrient forms that provide resources for other biota and primary production. In natural ecosystems, most N and a portion of the P required for plant growth are supplied through decomposition of organic matter, relying therefore on the activities of soil biota. Even in agricultural systems that typically receive large inputs of fertilizers, the supply of plant nutrients via the decomposition of organic matter is central to maintaining sustainable levels of crop production.

The aim of this chapter is to illustrate how the activities of soil biota, especially their trophic interactions, influence these processes of decomposition and nutrient cycling, and examine the significance of this for material flow and plant production in terrestrial ecosystems. The focus will be on the availability of N and P since they are the two nutrients that most limit primary productivity in natural and managed terrestrial ecosystems (Chapin et al. 2002). First, the issue of how soil microbes regulate the internal cycling of nutrients in terrestrial ecosystems will be discussed. This will be followed by a discussion of how soil animals influence nutrient cycling and plant growth through their feeding activities on microbes and other fauna. (The influence of plant-parasitic organisms on nutrient cycling will not be considered here, since this issue is discussed in more detail in Chapter 4).

The issue of how soil organisms and their interactions affect ecosystem properties by modifying the physical environment in which they live will then be discussed. Finally, a related and very topical issue that is high on the political and scientific agenda will be considered, that is, the question of whether variations in the diversity and architecture of soil food webs have important consequences for ecosystem functioning. In other words, is there a relationship between soil biological diversity and ecosystem functioning?

3.2   Microbial control of soil nutrient availability

In total, 80–90% of primary production enters the soil system as dead plant litter and roots, and the primary decomposers of this material are the bacteria and fungi. Initial fragmentation and ingestion of fresh organic material by detritivores increases its surface area for microbial colonization, but the microbes, via the production of enzymes, do most of the chemical alteration of this material. As noted in Chapter 2, microbes have the capacity to produce a vast array of enzymes that can degrade almost all plant-derived compounds.

In many ecosystems, fungi are the most abundant primary decomposers. They are better equipped than bacteria for carrying out the decaying of insoluble plant material since their hyphal networks can penetrate new substrates and proliferate both within and between dead plant cells. They can also produce a range of enzymes that are capable of degrading complex cell wall components, such as lignin, enabling them to gain access to more labile compounds that occur within the cell. Another advantage of fungi over bacteria is their ability to transport nutrients through their hyphal network to zones of exploitation. This means they have the capacity to exploit new substrates, even when nutrients such as N and P are limiting (Boddy 1999). In contrast, bacteria are either immobile or move passively through soil. Therefore, they tend to exhaust substrates that become available in their immediate microenvironment, and then become inactive until resources once more become available, for example, due to the exudation of soluble C compounds from newly growing roots.

3.2.1   Nitrogen mineralization

The mineralization of nutrients—the process by which soil microbes break down soluble and insoluble organic matter and convert it into inorganic forms—is of critical importance for ecosystem function since, in many ecosystems, it directly determines the availability of nutrients to plants. For example, in fertile ecosystems, such as deciduous forests, plant N supply is strongly correlated with rates of N mineralization in soil (Nadelhoffer et al. 1985) (Fig. 3.1). A summary of the soil N cycle is given in Fig. 3.2.

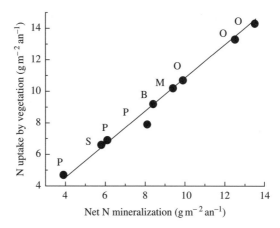

Fig. 3.1   Above-ground plant N uptake in relation to net N mineralization in the surface 0–20 cm soil. Symbols designate dominant genera on sites: O = oak, B = birch, M = maple, P = pine, S = spruce. (Data from Nadelhoffer et al. 1985.)

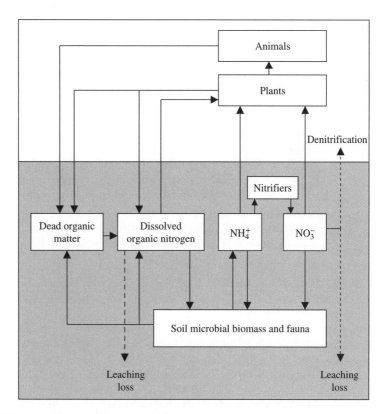

Fig. 3.2   Schematic diagram of the terrestrial N cycle.

Most N in the soil (some 96–98%) is contained in dead organic matter as complex insoluble polymers such as proteins, nucleic acids, and chitin. These polymers are too large to pass through microbial membranes, so microbes produce extracellular enzymes (e.g. proteinases, ribonucleases, and chitinases) that break them down into smaller, water soluble subunits that can be absorbed by microbial cells (e.g. amino acids). This material is called dissolved organic N (DON) and it can constitute a large proportion of the total soluble N pool, especially in infertile ecosystems with mor soils, such as Boreal forest and arctic tundra (Northup et al. 1995; Jones and Kielland 2002). Even in agricultural soils that receive regular dressings of inorganic N fertilizer, concentrations of DON in soil can be equal to, or even higher than, those of inorganic N (Bardgett et al. 2003).

The majority of DON in soil solution is absorbed by free-living soil microbes, which use the C and N contained within it for their growth. How microbes use DON, however, depends mainly on whether they are C- or N-limited. Under conditions when microbial growth is C-limited, microbes use the C from DON to support their energy needs for growth and they excrete plant available ammonium (NH_4^+) as a waste product into the soil; that is, N is mineralized by the microbial biomass. When DON is not sufficient to meet microbial N demand, however, microbes absorb additional inorganic N (NH_4^+ and NO_3^-) from soil solution; that is, N is immobilized by the microbial biomass, thereby reducing the availability of inorganic N for plant uptake. Besides plant uptake, ammonium produced by microbial mineralization has several potential fates: it can be absorbed onto negatively charged surfaces of clay minerals and organic matter and, when oxygen is plentiful, it can be oxidized by nitrifying bacteria to produce nitrate (Box 3.1).

Both immobilization and mineralization of N occur simultaneously in soil, but *net* N mineralization occurs when microbes are predominately C-limited, whereas *net* immobilization occurs when their growth is limited by N. The balance between mineralization and immobilization is determined by a range of factors such as the availability of DON and predation by soil animals, but one factor that is of high importance is the relative demand by microbes for N and C, which is determined by the C : N ratio of the organic substrate they are utilizing. Heterotrophic bacteria and fungi have C : N ratios ranging from 4 : 1 to 12 : 1, but as they break down organic matter they respire some 50% of the C contained within it and use the remaining 50% for biomass production (Kaye and Hart 1997). Therefore, there is a widely cited critical substrate C : N ratio of around 30 : 1 which is needed to meet microbial needs for N (Kaye and Hart 1997). If substrate C : N ratios are higher, which is often the case for plant litters, microbes become N-limited and hence use (or immobilize) exogenous sources of inorganic N while decomposing the substrate (Kaye and Hart 1997). In such situations, plants and microbes should theoretically compete for N in soil solution. Several

Box 3.1 Nitrification

Nitrification is the process by which ammonium (NH_4^+) is oxidized to nitrite (NO_2^-) and subsequently to nitrate (NO_3^-). This process is carried out by a restricted group of autotrophic bacteria that are classified into two groups on the basis of whether they oxidize NH_4^+ to NO_2^- (*Nitrosomonas* and other *Nitros-* genera) or NO_2^- to NO_3^- (*Nitrobacter* and other *Nitro-* genera). Heterotrophic nitrification is also known to occur, especially by fungi in forest soils. The process of nitrification is influenced by several soil factors, most notably the availability of NH_4^+, which is the sole energy source for autotrophic nitrifiers. Hence, rates of nitrification tend to be elevated in agricultural soils that receive large inputs of fertilizer N and in natural ecosystems where atmospheric N deposition is high. Oxygen is also crucial for nitrification to occur, so factors that influence the diffusion of O_2 through soil, such as moisture and soil structure, will also alter rates of nitrification in soil. Nitrification also varies with soil temperature, being maximal at 30–35°C, but it also known to occur in cold, but unfrozen soils under snow (Brooks et al. 1996). Rates of nitrification are also affected by soil moisture, and tend to be very low in dry soils, where thin water films restrict the diffusion of NH_4^+ to nitrifiers. Interestingly, nitrification is an acidifying process in that the oxidation of NH_4^+ to NO_3^- yields H^+ ions. However, it is also sensitive to changes in soil pH in that, in agricultural soils, nitrification is negligible below pH 4.5. While rates of nitrification tend to be quite low in temperate acid forest soils, it is elevated after disturbances, such as clear-cutting and fire that result in elevated rates of NH_4^+ production by mineralization.

Plants and microbes take up NO_3^- produced by nitrification, but significant amounts of NO_3^- may also be lost from soil, via leaching and denitrification. Compared to positively charged cations (e.g. NH_4^+) that are retained in soil by absorption to negatively charged soil colloids (e.g. clay minerals), NO_3^- is highly mobile in soil solution. Hence, when not used by plants or microbes, or when soils have a limited anion exchange capacity, it is readily lost from soil to drainage waters. This is especially the case in coarse textured, sandy soils and after rain events or when soils are irrigated. This leaching of NO_3^- is especially a problem in heavily fertilized agricultural soils where it often leads to pollution of groundwater, lakes, and streams. NO_3^- may also be lost from soil by denitrification, which is the microbial reduction of NO_3^- to gaseous NO, N_2O, and N_2. This process is carried out by a wide range of micro-organisms under conditions of high NO_3^- and C supply, and low oxygen, and is especially prevalent in wetlands that either receive inputs of NO_3^- from outside the systems or when they have a surface aerobic zone where nitrification occurs, such as in paddy fields. Denitrification is also prevalent in intensively managed grasslands in high rainfall regions, such as western Britain, where soils may be wetter than field capacity for much of the year. In these systems, the combination of high rainfall and fine textured soils compacted by intensive grazing, coupled with high NO_3^- from fertilizer application, creates conditions that are especially conducive to denitrification (Abbasi and Adams 2000). High rates of denitrification are of particular concern since it is a major source of N_2O, a greenhouse gas that contributes to the depletion of ozone in the stratosphere.

factors influence substrate quality in ecosystems, such as inherent soil fertility and plant community composition, and factors such as grazing that modify the quality of plant litter inputs to soil within ecosystems; the importance of these factors for soil nutrient cycling will be discussed in Chapters 4 and 5.

Not all DON in soil solution is used by microbes; significant quantities may be leached from soil in drainage waters (Perakis and Hedin 2002) and there is growing evidence that plants can uptake DON directly from soil in the form of amino acids, thereby bypassing microbial mineralization (reviewed by Lipson and Näsholm 2001). This has been shown to be the case in many ecosystem types, but is thought to be especially prevalent in less fertile ecosystems, such as the arctic tundra (Chapin et al. 1993; Kielland 1994; Schimel and Chapin 1996; Lipson and Monson 1998; Raab et al. 1999; Henry and Jefferies 2003) and Boreal forest (Näsholm et al. 1998; Nordin et al. 2001). In contrast, in agricultural situations, while plants clearly have the capacity to uptake amino acids directly (e.g. Näsholm et al. 2000, 2001; Streeter et al. 2000) this is thought to be of limited importance for plant nutrition due to very rapid microbial mineralization of DON (Hodge et al. 1998, 1999; Owen and Jones 2001; Bardgett et al. 2003). An exception to this, however, could be where amino acid availability is especially high, for example, in resource hot spots or patches in soil. Many plants have the capacity to exploit such resource patches by root proliferation, and hence, in these situations, the acquisition of organic N could be a significant pathway of plant nutrition (Hodge et al. 1998).

3.2.2 Nitrogen fixation

Another route through which plants gain N is via the process of N fixation. This is especially the case in natural, unpolluted terrestrial ecosystems where N fixation is often the primary route by which N enters the system (Cleveland et al. 1999). Microbes that have the capacity to reduce molecular N_2 to NH_3, and incorporate it into amino acids for protein synthesis, drive this process of N fixation. These microbes are either free-living or

Table 3.1 Rhizobia and their host plants. (Adapted from Paul and Clark 1996)

Genus	Species	Host plant
Rhizobium	*meliloti*	Alfalfa (*Medicago, Melitotus*)
	leguminosarum	Peas (*Pisum*)
		Vetches (*Vicia*)
		Clovers (*Trifolium*)
		Beans (*Phaseolus*)
	loti	Trefoil (*Lotus*)
	fredii	Soyabean (*Glycine*)
Bradyrhizobium	*japoniucum*	Soyabean (*Glycine*)
		Tropical legumes (*Arachis, Leucaena*)
Azorhizobium	*caulinodans* spp.	Stem nodules (*Sesbania*)
		Non-legumes (*Parasponia*)

they live in symbiotic association with plants, forming nodules in the root where they receive carbohydrate from the plant to meet their energy needs and in turn they supply the plant with amino acids formed from reduced N.

The most widely known N fixers are probably the legumes that are associated with bacterium of the genus *Rhizobium* (Table 3.1) and non-legumes such as alder that have the actinomycete *Frankia* as their endophyte (Table 3.2). These symbiotic associations can supply large amounts of N into soil of both natural and agricultural ecosystems, and their benefits for soil fertility have been recognized since pre-Roman times. In agricultural grassland, for example, legumes can contribute up to 150 kg N ha^{-1} yr^{-1} (Newbould 1982), and in natural situations, especially in early primary succession, they are often the major contributors to N accumulation (Walker 1993). The primary route by which this N enters the soil from legumes is through the breakdown of litter inputs that are enriched with N. Another point of entry is via the exudation or leakage of N-rich exudates from roots.

Table 3.2 Distribution of actinorhizal plants. (Adapted from Paul and Clark 1996)

Continent	Native genera
North America	*Alnus, Ceanothus, Cerocarpus, Chamaebatia, Comptonia, Coraria, Cowania, Datisca, Dryas, Elaeagnus, Myrica, Purshia, Shepherdia*
South America	*Alnus, Colletia, Coriaria, Discaria, Kentrothamnus, Myrica, Retanilla, Talguenea, Trevoa*
Africa	*Myrica*
Eurasia	*Alnus, Coriaria, Datisca, Dryas, Eleagnus, Hippophae, Myrica*
Oceania, including Australia	*Allocasuarina, Casuarina, Ceuthostoma, Coriaria, Discaria, Gymnostoma, Myrica*

Details on the establishment of symbiosis, the biochemistry of N fixation, and the factors that influence rates of N fixation, are given in Paul and Clark (1996).

Another potentially important pathway of N fixation is through free-living bacteria that can fix N in the course of decomposing litter and soil organic matter (Vitousek and Hobbie 2000). These non-symbiotic N-fixers are ubiquitous in terrestrial ecosystems (Paul and Clark 1996), and fix relatively small, but significant amounts of N (<3 kg N ha^{-1} yr^{-1}) (Cleveland et al. 1999). In extreme environments with sparse cover of vascular plants, cyanobacteria are often the main source of N. These N-fixing organisms are especially abundant in dry and/or cold regions, where the growth of vascular plants is limited. For example, in arid ecosystems they occur in soil crusts (Box 3.2) composed of cyanobacteria, lichens, and mosses (Belnap 2003). Similarly, cyanobacteria are abundant as crusts or felts on the soil in the Antarctic (Wynn-Williams 1993) and in recently deglaciated terrain, such as that found at Glacier Bay (Fig. 3.3) (Chapin et al. 1994). In all these situations, they are vital for creating and maintaining fertility, often fixing appreciable amounts of N ($1–10$ kg N ha^{-1} yr^{-1}) (Vitousek et al. 2002); they also have an important role in stabilizing surface soil particles in extreme environments (Wynn-Williams 1993; Belnap 2003).

Recent findings also point to cyanobacteria as being the primary N source in Boreal forest, where N fixation has traditionally been thought to be extremely limited. DeLuca and colleagues (2002) showed that a N-fixing symbiosis between a cyanobacterium (*Nostoc* sp.) and the ubiquitous feather moss *Pleurozium schreberi* fixes significant quantities of N ($1.5–2.0$ kg N ha^{-1} yr^{-1}), and acts as a major contributor to N accumulation and cycling in Boreal forests (Fig. 3.4). This finding is especially significant since *P. schreberi* is the most common moss on Earth, found throughout North and South America, Greenland, Asia, Europe, and Africa, and in Boreal forests, it often forms a continuous carpet, accounting for up to 80% of the ground cover (DeLuca et al. 2002).

3.2.3 Microbial phosphorus mineralization

In contrast to N, most P in soil is in a range of insoluble inorganic forms that are unavailable for plant uptake, for example, as occluded P (Fig. 3.5). As a result, many of the reactions that govern plant availability of P are geochemical, rather than biological (Box 3.3). Soil P availability is also strongly affected by past management. For example, P fertilization has a substantial and incredibly long-term effect on soil P availability, in that it greatly elevates plant available phosphate (PO_4^{3-}) concentrations in soil, especially in surface horizons. Despite this, soil microbes are closely involved in the cycling of P, in that they participate in the solubilization of inorganic P and

Box 3.2   Biological soil crusts

Extreme environments such as deserts are often thought of as being life-less places. However, despite their barren appearance, they are teaming with life; rocks and sand on the soil surface are often covered with a thin film of cyanobacteria, microfungi, lichens, and/or mosses, collectively known as 'biological soil crusts'. Most of the cohesion in biological crusts is due to large filamentous cyanobacteria, whose filaments link other-wise loose particles together to form larger soil aggregates. Key to their survival in extreme environments is their ability to tolerate prolonged periods of dehydration, resulting from lengthy periods of high heat, strong light, and no water (Belnap 2003). Also, many species have evolved ways of prolonging their activity; for example, many lichens can photo-synthesize over a wide range of temperatures while others require little water to begin metabolism (Lange 2001). Pigments are also used by many crust species to reflect and/or absorb excessive radiation (Belnap 2003).

Biological crusts have many roles in the formation, stabilization, and fer-tility of soils (Belnap 2003). They capture nutrient-rich dusts, often increasing concentrations of plant growth, limiting nutrients in soil (Reynolds et al. 2001), increasing both the fertility and water-holding capacity of soils (Verrechia et al. 1995). Crusts also add nutrients to soil via fixation of atmospheric N and C; in deserts, C inputs from cyano-bacterial crusts range from 0.4 to 2.3 g m^{-2} (Evans and Lange 2001) and N inputs from N fixation are typically around 1 kg ha^{-1} yr^{-1} (Belnap 2002), representing the dominant source of N (Evans and Ehleringer 1993). They also enhance soil fertility by exuding negatively charged polysaccharides that increase soil aggregation and retain positively charged nutrients (e.g. NH_4 and Ca^{2+}), thereby preventing their leach-ing from soil. Soil food webs beneath crusts are also often more abund-ant and diverse than in the surrounding environment, thereby further promoting rates of decomposition and nutrient cycling. All these factors contribute to increased soil fertility, often promoting the growth and nutritional status of vascular plants. These functionally important soil organisms are extremely vulnerable to climate change and disturbances such as off-road vehicles and grazing livestock; further, their recovery from such disturbances is extremely slow, taking decades to centuries.

in the mineralization of organic P, the latter being governed by the pro-duction of plant and microbial phosphatases in situations of low P avail-ability. These enzymes act by cleaving ester bonds in organic matter to liberate phosphate (PO_4^{3-}) that can be taken up by the plant or microbial biomass.

Fig. 3.3   Black algal crusts (left) adjacent to the pioneer N-fixer *Dryas drummondii* (right) that colonize recently deglaciated terrain in Glacier Bay, Alaska. (Image by Richard Bardgett.)

Fig. 3.4   Carpets of feather moss *Pleurozium schreberi* develop on the floor of Boreal forests, such as those in Alaska, where in association with cyanobacteria it fixes significant quantities of N, and acts as a major contributor to N accumulation and cycling. (Image by Richard Bardgett.)

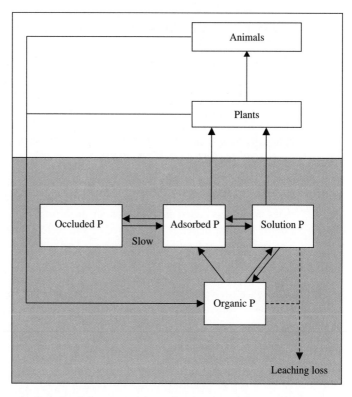

Fig. 3.5   Schematic diagram of the terrestrial P cycle, showing main transformations of P that occur in soil.

As with N, the mineralization of organic P is partly regulated by the C : P ratio of substrates. In general, when the C : P ratio rises above ~100, P is immobilized by microbes which have a relatively high P requirement (1.5–2.5% P by dry weight compared to 0.05–0.5% for plants). As a result, microbes compete aggressively with plants for available P in soil. The importance of microbial immobilization of P is illustrated by the fact that microbes often contain as much as 20–30% of the total soil organic P pool (Jonasson et al. 1999*a*), which is much larger than the proportion of C (~1–2%) or N (~2–10%) held in microbes. Also the capacity of the soil microbial biomass to act as a sink for P is strongly affected by environmental stresses, such as wetting and drying cycles, that are known to result in the death of a portion of the soil microbial biomass and a resultant flush of available P, and presumably other nutrients, into soil solution (Turner and Haygarth 2001). Microbial P is, therefore, an important source of potentially available P in soil. Mycorrhizal fungi also play an important role in the transfer of P in soil systems, as will be discussed in the next section.

Box 3.3   Phosphate availability

Phosphate is the main inorganic form of P that is available to plants. This phosphate is often divided into two forms: (1) The *labile* P pool, which constitutes the P on mineral surfaces that can be easily desorbed by plants and phosphate ions in soil solution. This form of P is immediately available to plants; and (2) P that precipitates with Fe and Al compounds, and becomes insoluble and unavailable to plants. This latter form of P is referred to as *occluded* P.

Amounts of labile P in soil are very low relative to the total P pool, because any P released by mineralization is rapidly adsorbed by soil particles, a process called *fixation*, rendering it unavailable to plants. The rate of fixation depends on a range of soil factors, but mostly soil pH. The most serious fixation occurs in very acid soils (pH <5) where there are large amounts of Fe and Al hydroxides in soil solution. These combine with phosphate to produce Fe and Al phosphates that are insoluble and hence unavailable to plants. This is especially prevalent in highly weathered soils, such as those of the tropics (i.e. Oxisols and Ultisols) where the growth of plants is often strongly P limited. In calcareous soils, where pH values exceed 7, a different type of fixation occurs. Here, phosphate combines with calcium to form insoluble phosphates such as tricalcium phosphate $(Ca_3(PO_4)_2)$. Fixation of P is at its lowest in soils of pH 5.5–6.5, since fixation by Fe, Al, and Ca is minimal. Hence, availability of P to plants is maximal at this pH range.

Attention is turning to the role of organic P in soils, especially inositol phosphates that are the dominant class or organic P compounds in soil (Turner et al. 2002). A large proportion of the total soil P is present in organic forms, typically between 29% and 65% (Harrison 1987). The majority of the identifiable soil organic P exists as orthophosphate monesters (60–90% of the organic P), and inositol phosphates constitutes some 60% of this pool (Turner et al. 2002). Inositol phosphates are derived from plants and exist in soil in various forms. Despite their widespread occurrence in soils, their role in P cycling remains poorly understood. Particular attention is been placed on understanding the factors that regulate its availability in soil, and its potential to provide plants with P in P-limited ecosystems. There is also much interest in its transfer from terrestrial to aquatic-ecsosystems, and its potential to influence water quality. The reader is referred to Turner et al. (2002, 2004a) for further information on the nature and importance of organic P in terrestrial and aquatic ecosystems.

3.2.4 The role of mycorrhizal fungi in plant nutrient supply

Mycorrhizae are distributed across a wide range of ecosystems, but there is a distinct pattern in the distribution of different mycorrhizal types according to biome, soil type, and limiting resources (Fig. 3.6) (Read 1983). In general, ericoid mycorrhizal fungi dominate in high latitude and altitude ecosystems, which are dominated by dwarf-shrub heath vegetation on peaty N-limited soils; Ectomycorrizal fungi dominate Boreal and broad-leaved forest ecosystems, that typically occur on brown earth and podzolic soils; whereas vegetation of tropical forest, grassland, and desert predominately have AM fungal associations. This pattern provides a simple framework to characterize the functional role of mycorrhizae in relation to plant community types.

The primary benefit to plants of having mycorrhizal associations is enhanced acquisition of nutrients from soil, especially of N and P. Indeed, mycorrhizal fungi often constitutes one of the largest C sinks for primary productivity (Harley 1971), and allocation to these fungi is especially high when nutrients are limiting (Johnson et al. 2003*c*). It has long been known that mycorrhizal fungi enhance ammonium uptake from soil, and more recently it has been shown that both ericoid and ectomycorrhizal fungi have the capacity to uptake DON directly from soil, in the form of amino acids, pure protein, and even as highly recalcitrant proteins that are co-precipitated with tannins (Read 1994). They can use these more complex proteins

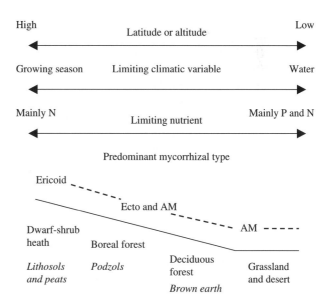

Fig. 3.6	Relationships between habitat types, limiting resources, and mycorrhizal associations. (Adapted from Read 1983.)	

because they produce carboxyl-proteinase enzymes that have the ability to cleave proteins into constituent amino acids. Interestingly, the production and activity of these enzymes is optimal at soil pHs of 2–4, a range that coincides with that of the mor soils where these fungi typically occur (Leake and Read 1990). There is also emerging evidence that AM fungi can both enhance decomposition of, and increase N capture from, complex organic material in soil (Hodge et al. 2001); the significance of this for the growth and nutrition of AM plants is yet to be established.

Another route by which mycorrhizal fungi can influence soil N availability is by increasing the fixation of N by plant-microbial N-fixing associations. AM fungal infection, for example, is known to increase rates of nodulation and N fixation in legumes (Haystead et al. 1988) and in plants with acti-norhizal associations (Gardner et al. 1984), thereby potentially increasing the supply of N to soil. AM fungi can also transport fixed N from legumes, through soil, to neighbouring plants via hyphal networks. For example, N transfer between agricultural legumes and grasses has been shown to occur via AM hyphal networks (e.g. Haystead et al. 1988; Zhu et al. 2000) and it has been proposed—but not tested—that this route is of greatest significance in N-limited, early successional plant communities where legumes are often strongly mycorrhizal (Haystead et al. 1988). Similar transfers of N are also thought to occur via ectomycorrhizal fungi between actinorhizal alder (*Alnus sinuata*) and its later successional neighbours (Allen 1991).

Mycorrhizal fungi also have an important role to play in the uptake of soil P by host plants. These fungi grow deep into the soil matrix, accessing soil P that is beyond the roots. They do this through two important mechanisms. First, they are known to produce phosphatase enzymes that cleave ester bonds that bind P to C in organic matter, thereby releasing phosphate (PO_4^{3-}) that can be taken up by the fungi and passed on to the plant. Second, they produce low molecular weight organic acids, such as oxalates, which enhance the availability of soil P by increasing weathering rates of P contained in clay minerals, and by complexing cations (e.g. Ca, Fe, and Al) that would otherwise bind to phosphates, thereby taking them out of soil solution.

3.3 Influence of animal–microbial interactions on nutrient availability

Mineralization of nutrients is governed directly by the activities of bacteria and fungi. The ability of microbes to do this, however, is affected strongly by soil animals that live alongside them, and also by food web interactions that determine the transfer of nutrients through the plant-soil system. There are many ways that soil animals can affect microbial nutrient mineralization, but for convenience they are typically divided into three broad pathways. First, soil animals affect nutrient cycling through their selective feeding on microbes, which alters microbial activity, abundance, and community structure. Second, soil animals affect nutrient mineralization by altering the form of the organic matter in soil, in that they fragment and mix organic matter inputs to soil, increasing its susceptibility to microbial attack. Third, it has been suggested that soil animals, notably protozoa, can have non-nutritional effects on plant growth, in the form of hormonal effects on root morphology. Combined, these interactions between microbes and animals drive processes of energy flow and nutrient cycling, and, therefore contribute to plant nutrient acquisition and plant growth. In the following section, the effects of these biotic interactions on the structure and activity of microbial communities, and the consequences of this with regard to nutrient availability to plants will be discussed.

3.3.1   Selective feeding on microbes by soil animals

While soil animals are increasingly considered to be both generalist and opportunist in their feeding behaviour (Scheu and Falca 2000; Ponsard and Arditi 2000; Maraun et al. 2003), there is also evidence that many of them are very selective in this regard. For example, it is well known that certain nematodes have mouth parts that are adapted to feeding on either bacteria (bacterivores) or fungi (fungivores) (Yeates et al. 1993), and there is evidence that certain protozoa use chemical cues to discriminate between bacterial and algal prey (Verity 1991). Other faunal groups, notably the collembolans, have been shown to be even more selective in their feeding behaviour, actually choosing to feed on particular fungal species over others (e.g. Bardgett et al. 1993*b*). One of the best examples of this is that of Newell (1984*a,b*) who showed that selective grazing by the collembolan *Onychiurus latus* on the fungus *Marasmius androsaceus* in coniferous leaf litter resulted in a reduction in the activity of this palatable fungus, and an increase in the abundance of an unpalatable fungus *Mycena galopus* present in the litter. This change in fungal community structure then had a knock-on effect at the ecosystem scale, in that it reduced decomposition rates of coniferous litter, since *M. galopus* decomposes litter at a slower rate than *M. androsaceus*. This kind of selection of prey can be based on numerous factors, but it is mostly related to the palatability of the fungus, which varies with species, age, and physiological status. For example, the collembolan *Folsomia candida* has been shown to feed on metabolically active hyphae in preference to dead or inactive hyphae (Moore et al. 1985) and also selects regions of the fungal thallus with high N content (Leonard 1984). Preferential grazing is also thought to arise as a result of the avoidance of toxins that are produced by some fungal colonies (Parkinson et al. 1979), and some species of Collembola have been shown to locate and select their fungal food source by volatile compounds released from fungal mycelium (Bengtsson et al. 1988).

Preferential grazing by collembolans on fungus has dramatic effects on the activity and abundance of the prey, and, as shown above, can also alter fungal community structure. It is well known, for example, that an intermediate level of grazing by collembolans actually enhances the activity and growth of selected fungi (Hanlon and Anderson 1979; Bengtsson and Rundgren 1983; Bardgett et al. 1993c). This stimulation of fungal growth, often referred to as compensatory growth, is due to new fungal growth after senescent hyphae are grazed, and regrowth after periodic grazing of actively growing mycelia. For example, a study by Hedlund et al. (1991) showed that grazing of the fungus *Mortierella isabellina* by the collembolan *O. armatus* induced switching from a 'normal' hyphal mode, with appressed growth and sporulating hyphae, to fan-shaped sectors of fast growing and non-sporulating mycelium with extensive areas of aerial hyphae. These authors also showed that the activities of specific amylase enzymes were several times greater in grazed cultures than in those cultures that were ungrazed. In contrast, heavy grazing of fungal communities can reduce their activity, as shown by Warnock et al. (1982) who found that heavy grazing of mycorrhizal fungi by collembolans counteracted their mutualistic relationship with the host plant.

The ingestion and exposure of microbes to intestinal fluids can greatly influence microbial activity. For example, microbial communities have been shown to be more abundant and active following passage through the gut of earthworms (Brown 1995), and dormant bacteria can be activated during this intestinal journey due to the removal of endospore cell coats and subsequent germination of bacterial spores (Fischer et al. 1997). However, the earthworm gut can also be a hostile environment for microbes. As shown by Moody et al. (1996), the germination of fungal spores can be reduced following ingestion by earthworms, due to exposure to intestinal fluids.

3.3.2 Effects of microbial-feeding fauna on nutrient cycling and plant growth

The effects of microbial-feeding microfauna on microbial activity, nutrient mineralization, and primary production are generally positive (Mikola et al. 2002). Enhanced C mineralization results from increased turnover rate, activity, and respiration of grazed microbial populations (Anderson et al. 1981; Kuikman et al. 1990; Bardgett et al. 1993c; Mikola and Setälä 1998a; Cole et al. 2000), whereas enhanced N mineralization is mainly due to direct animal excretion of excess N (Woods et al. 1982). In general, grazers have lower assimilation efficiencies than the microbes upon which they graze, and therefore they excrete into soil nutrients that are not required for production, in forms that are biologically available (e.g. protozoa preying on bacterial populations are assumed to release about one-third of the N consumed). This release of nutrients into the soil system is effectively a remobilization of the nutrients that were bound up in the microbial biomass, and has been termed the 'microbial loop' (Clarholm 1985).

The significance of the 'microbial loop' is that the nutrients released from microbial biomass by grazers increase the availability and uptake of nutrients by plants (Clarholm 1985), and, in some cases, enhance plant growth. For example, a classic study by Ingham et al. (1985) showed that the addition of microbial-feeding nematodes to soil in pot experiments enhanced N uptake by, and growth of, the grass *Bouteloua gracilis* (Fig. 3.7(a)). Similarly, Bardgett and Chan (1999) showed that, when acting together, Collembola and nematodes increased concentrations of ammonium in the soil solution of an upland organic soil, leading to increased plant nutrient uptake by, but not growth of, the grass *Nardus stricta* (Fig. 3.8). Studies of trees also show similar responses to the addition of animals. For example, Setälä and Huhta (1991) showed that leaf, stem, and shoot biomass of birch seedlings (*Betula pendula*) were increased when the plants were grown in the presence of a diverse soil fauna (Fig. 3.7(b)), and Setälä (1995) showed that grazing of ectomycorrhizal fungi associated with *B. pendula* by soil fauna resulted in enhanced growth of, and nutrient uptake by, *B. pendula*, despite reductions in the biomass of ectomycorrhiza. There is therefore ample evidence to suggest that faunal grazing on microbes can enhance both microbial activity and the availability of nutrients for plants, thereby influencing net primary productivity (NPP).

Evidence is also emerging that interactions between decomposer organisms can influence the performance of higher trophic groups that live aboveground, especially herbivores that benefit from enhanced plant nutrition. For example, Scheu et al. (1999) showed that plant-sucking aphids

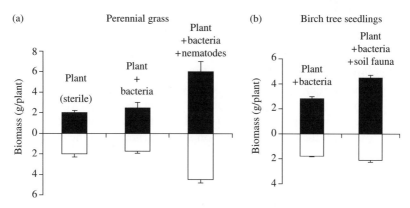

Fig. 3.7   Influence of soil trophic interactions on plant growth. (a) Adding bacterial-feeding nematodes to soil increased shoot and root growth of the perennial grass *Bouteloua gracilis* (data from Ingham et al. 1985). (b) Addition of diverse fauna to soil increased root and shoot biomass of birch seedlings (*Betula pendula*). Filled bars = shoot biomass; open bars = root biomass. (Data from Setälä and Huhta 1991.)

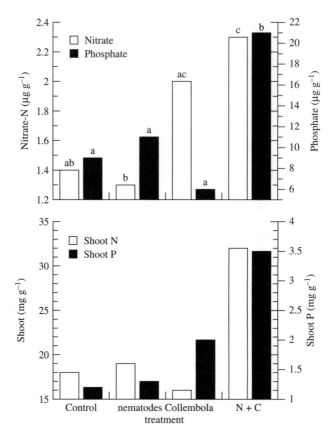

Fig. 3.8 Effects of animal treatments (N = nematodes, C = Collembola, and N + C = nematodes and Collembola in combination) on the amount of NO_3^-–N ($\mu g\ g^{-1}$ soil) and PO_4^{3-}–P ($\mu g\ g^{-1}$ soil), leached from microcosms, and shoot N and P content (mg g^{-1}). Values are means $\pm$SE. For each measure, values with the same letter are not significantly different. (Data from Bardgett and Chan 1999.)

performed better where the host plant was grown in the presence of microbial-feeding Collembola or earthworms, as compared to systems without these organisms. Similarly, bacterial-feeding microfauna were shown to indirectly increase the numbers and biomass of aphids on barley shoots through their positive effects on soil N turnover and the nutritional status of the plant (Bonkowski et al. 2001a), and the presence of earthworms has been shown to increase foliage consumption by a leaf chewer (*Mamestra brassicae*), due to a positive effect of these soil animals on soil N availability and foliar N content (Newington et al. 2004). It is therefore apparent that important indirect linkages and feedbacks operate between the consumer organisms of the above-ground and below-ground food webs.

It is important to recognize that effects of soil animals on nutrient mineralization and plant nutrient uptake are case specific. For example, soil fauna have been shown to both reduce or enhance microbial activity and mineralization depending on the season (Teuben 1991) and the abundance of grazers (Hanlon and Anderson 1979; Hanlon 1981). The effects of interactions between fauna and microbes on nutrient release are also highly dependant on resource quality. In N-limited soils, for example, nutrients released by fauna grazing on fungi are often rapidly re-utilized by microbes, and hence do not become available for plant uptake (Visser et al. 1981; Bardgett et al. 1993*c*; Cole et al. 2002*b*). This suggests that in nutrient limited conditions, fungal-feeding animals alone are not able to limit the ability of microbes, and especially fungi, to sequester nutrients and hence compete with plants. Effects of animal–microbial interactions on nutrient cycling are also likely to be spatially constrained, occurring largely in hot spots of activity around resource patches (Bonkowski et al. 2000).

3.3.3 Non-nutritional effects of microbial grazers on plant growth

Beneficial effects of protozoa on plant growth are well documented (Clarholm 1985; Kuikman et al. 1990; Jentschke et al. 1995; Alphei et al. 1996; Bonkowski et al. 2000), and are usually assigned to nutritional effects via the 'microbial loop'. In recent years, however, there is ample evidence to suggest that protozoan grazing could also have non-nutritional effects on plant growth. The idea here is that protozoa indirectly influence root architecture by influencing bacterial production of plant growth-promoting hormones. A number of studies show that plants grown in the presence of protozoa develop an extensive and highly branched root system, due to strong branching of lateral roots, and that these effects resemble those caused by plant growth-promoting hormones, such as auxins (Jentschke et al. 1995; Bonkowski et al. 2000, 2001*b*; Bonkowski and Brandt 2002). While it has long been known that protozoa can themselves release auxins (Nikolyuk and Tapilskaja 1969), recent studies reveal that selective grazing by protozoa in the rhizosphere significantly stimulates the growth of auxin-producing bacteria; this has been proposed as the most likely mechanism of protozoan effect on root growth (Bonkowski and Brandt 2002; Bonkowski 2004) (Fig. 3.9). Changes in root architecture may also be due to other effects of protozoa grazing on the bacterial community. In particular, stimulation of nitrifying bacteria may lead to hot spots of nitrate concentration; as well as being a source of N for plant growth, nitrate acts as a signal for lateral root elongation and may act to direct root growth towards nutrient patches (Zhang and Forde 1998). Overall, these studies reveal that protozoan effects on plant growth are more complex than previously assumed, in that a combination of nutritional and non-nutritional responses to microbial grazing are at play (Bonkowski 2004).

3.3.4 Multitrophic controls on soil processes

Effects of biotic interactions in soil on decomposition and nutrient release are complex and involve a diversity of species from more than one trophic

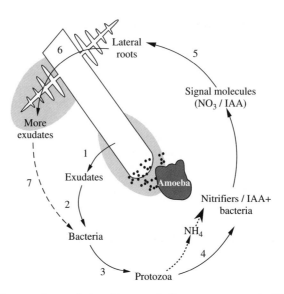

Fig. 3.9   Schematic diagram of grazer-induced hormonal effects on root growth. (1) plant roots release exudates thereby: (2) stimulating growth of bacteria; and (3) bacterial predators; (4) selective grazing by protozoa favours auxin-producing and nitrifying bacteria; (5) inducing lateral root growth; (6) leading to enhanced exudation; and (7) bacterial growth. (Redrawn with permission from the New Phytologist Trust; Bonkowski 2004)

group. For example, the effects of feeding of fungal-feeding Collembola on nutrient cycling have been shown to become apparent only when they are interacting with another trophic group of soil fauna, namely microbial-feeding nematodes (Bardgett and Chan 1999). Likewise, it has been demonstrated that combinations of soil animals, as opposed to a single group of soil fauna, had a synergistic effect on the microbial community in microcosms of coniferous forest humus, resulting in enhanced leaching of mineral nutrients (Setälä et al. 1991). Animals also prey on each other which complicates the interpretation of how soil fauna affects ecosystem processes. Not much is known about the importance of predation in soil food webs, but the studies to date reveal that manipulation of predatory fauna can have dramatic effects on processes of decomposition and nutrient mineralization and that these effects can be both positive and negative, depending on the particular circumstances. This occurs because predatory fauna can induce trophic cascades that lead to positive, neutral, and negative effects on the activity or biomass of microbes. For example, Santos et al. (1981) found that reductions in predatory mites in soil food webs of desert soils led to increased abundance of their prey, bacterial-feeding nematodes, which in turn led to reduced microbial growth and decomposition of plant litter. In other studies, however, reductions in abundance of microbial-feeding nematodes by predation have led to neutral (Laakso and Setälä 1999a) or negative (Bouwman et al. 1994; Mikola and Setälä 1998a,b; Setälä et al. 1999) effects on nutrient mineralization, presumably due to differential response of microbes to changes in their predation. In sum, it is clear that predatory fauna have the potential to induce cascading effects on soil food webs that ultimately influence rates of nutrient mineralization and hence plant nutrient availability.

3.4   Effects of animals on biophysical properties of soil

Soil fauna also affect ecosystem processes of nutrient cycling via physical alteration of decomposing material and the soil environment. Two groups of organisms are responsible for this: (1) the litter transformers that consume plant detritus and egest this material into soil as fecal pellets, thereby affecting rates of decomposition and nutrient release; and (2) the ecosystem engineers that build physical structures in soil that provide habitats for microbes and other organisms, and also alter the movement of materials through soils and across ecosystems (Lavelle et al. 1995).

3.4.1   Consumption of litter and the production of fecal pellets

Litter consumers are animals such as microarthropods and some macrofauna that consume plant detritus and egest this material into soil as fecal pellets. Such fecal pellets have a higher surface-to-volume ratio compared to the original leaf litter, which enhances its rate of decomposition (Webb 1977). Fecal pellets also provide a highly favourable environment for microbial growth, especially of bacteria, again leading to increased rates of decomposition and nutrient release (Hassall et al. 1987; Zimmer and Topp 2002). Decomposition is also enhanced through the action of endosymbiotic microbes that reside in the guts of many soil animals, such as earthworms, isopods, and termites. These endosymbiotic microbes produce extracellular enzymes that degrade cellulose and phenolic compounds thereby further enhancing the degradation of ingested material (Zimmer and Topp 1998).

Earthworms consume vast amounts of litter fall in many ecosystems, but especially those in mull type soils (Table 3.3). Indeed, in mixed forest ecosystems, earthworms can consume the entire annual litter fall of the forest (Nielson and Hole 1964; Satchell 1967). This organic matter is consumed along with soil mineral particles, and these two fractions are mixed together in the earthworm gut and then egested as surface or subsurface casts. Estimates from various habitats around the globe show that earthworms produce between 2 and 250 ton of cast ha^{-1}, whereas typical values for many temperate ecosystems probably range from 20 to 40 ton per hectare (Edwards and Bohlen 1996). It is widely documented that these casts contain significantly greater numbers of microbes and have higher

Table 3.3 Amount of organic matter ingested or incorporated into soil by earthworm populations of different habitats (Adapted from Bohlen 2002.)

Ecosystem	Type of organic matter	Amount consumed or incorporated (kg ha^{-1} yr^{-1})
Maize field	Maize residues	840
Orchard	Apples leaves	2000
Mixed forest	Canopy tree leaves	3000
Oak forest	Oak leaves	1071
Alfalfa field	Alfalfa residues	1220
Tallgrass prairie	Total organic matter	740–8980
Savanna	Total organic matter	1300

enzyme activities than the surrounding soil, either because of enrichment of organic matter with microbes in the worm's intestine or because the cast provides a better substrate for microbial growth, being rich in organic matter and available nutrients. As a result, rates of decomposition and C mineralization in soil are significantly enhanced by the presence of earthworms (Cortez et al. 1989). Similarly, rates of N mineralization have been found to be greater in earthworm casts relative to surrounding soil, largely due to them being enriched with inorganic N and N-rich excretory products and mucus from the earthworm (Lavelle and Martin 1992). P availability is also reported to be much greater in casts than surrounding soil, largely due to the stimulation of phosphatase activity (Sharpley and Syers 1976). While these positive effects of casting on microbial activity are often transitory (Lavelle and Martin 1992), their net effect, along with other earthworm activities, is the stimulation of total soil nutrient availability (Anderson et al. 1983; Scheu 1987) and enhanced plant nutrient uptake (Edwards and Bohlen 1996).

3.4.2 Physical engineering of the soil structure

Soil fauna also affect ecosystem properties by acting as ecosystem engineers (Lavelle et al. 1997), substantially modifying the physical structure of the soil profile, and hence the habitat and activities of other organisms and the passage of materials through soils. Engineers tend to be macrofauna, which exploit plant litter at the soil surface, such as earthworms and termites, and create macropores and channels as a consequence of their feeding and burrowing activities. Macrofauna also improve soil porosity and drainage through their burrowing activities. For example, one of the main effects of earthworms on soil porosity is to increase the proportion of macropores and channels in soil (Knight et al. 1992; Lavelle et al. 1997). These structures collectively act to enhance the movement of both water and soluble nutrients through soil and to interconnected waterways. Indeed, studies that have eliminated earthworms from pasture, through the use of pesticides, have observed dramatic decreases in water infiltration of up to 93% (Sharpley et al. 1979). Such effects of earthworms on infiltration, especially by deep-burrowing species, are known to increase leaching of nutrients from soil to groundwaters. Studies of grain-crop agro-ecosystems, for example, showed that the inoculation of soils with earthworms produced a 4- to 12-fold increase in leachate volumes and a 10-fold increase in N contained within them (Subler et al. 1997). Similarly, the inoculation of earthworms (*Aporrectodea caliginosa*) into limed coniferous forest soils produced a 50-fold increase in the concentration of nitrate and cations in soil solution (Robinson et al. 1992, 1996).

Nests of soil-nesting ants provide extensive macroporosity to the soil that affects infiltration rates of water and soluble nutrients. Bulk flow along nest galleries also provides an important route for recharge of deep soil moisture in arid and semi-arid environments. For example, in semi-arid Western Australia, ant biopores were found to transmit water down the soil profile when the soil was saturated and water was ponding on the surface (Lobry de Bruyn and Conacher 1994). Similarly, water infiltration into soils with nest entrances of funnel ants (*Aphaenogaster barbigula*), which can reach densities of up to 88,000 ha^{-1}, averaged 23.3 mm min^{-1} but was only 5.9 mm min^{-1} in nest entrance free soil (Eldridge 1993). Although not quantified, the above effects of ants on water infiltration are likely to have strong influences on transfers of water and nutrients to groundwaters, and also on the movement of water and nutrients to adjacent ecosystems (Bardgett et al. 2001*a*).

Soil fauna can also indirectly affect the hydrology of ecosystems by altering rates of organic matter accumulation. A fine example of this comes from subantarctic Marion Island, where the soil-borne larvae of a flightless moth, *Pringleohaga marioni*, process some 1.5 kg plant litter or peat m^{-2} yr^{-1}, thereby stimulating N and P mineralization 10-fold and 3-fold, respectively (Smith and Steenkamp 1990). Recently, the house mouse, *Mus musculus*, was introduced onto the island, which feeds on the moth larvae, thereby reducing annual litter turnover by some 60% of that originally processed by the moth (Smith and Steenkamp 1990). According to Smith and Steenkamp (1990), if mouse numbers continue to increase, rates of peat accumulation will increase, which in turn will dramatically alter the hydrological regime and vegetation of the island.

3.5 Functional consequences of biological diversity in soil

In recent years, there has been an explosion of studies that examine the ecosystem consequences of declines or changes in biological diversity, and these studies of diversity–function relationships have spurred considerable debate: one group of workers report that ecosystem productivity is positively related to species diversity, whereas the other group argues that there is no relationship between biodiversity and productivity, and that ecosystem function is explained by species identity rather than diversity *per se* (i.e. the absolute number of species with the community) (see Loreau et al. 2002). Most studies on diversity–function relationships, however, have been done on plant and aquatic communities, with the issue of how soil biological diversity affects ecosystem properties receiving only minor attention. This section considers the potential importance of soil biological diversity, and asks whether changes in the diversity of soil animal communities could significantly affect ecosystem properties.

A very common view amongst soil ecologists is that since species richness in soil is so high (Chapter 2) and there are a large number of trophically equivalent organisms, most species must be functionally redundant, that is, they are replaceable with other species without influencing general soil functions, such as nutrient and C mineralization (Andrén et al. 1995; Lawton et al. 1996; Groffman and Bohlen 1999; Setälä et al. 2005). In other words, the loss of individual species from the soil community will not result in a change in ecosystem properties, since another species will replace its role. An opposing view, however, is that this redundancy in soil is largely assumed without evidence (Behan-Pelletier and Newton 1999) and that within trophic groups differences among species are sufficient to produce species-specific impacts on ecosystem functioning (Wall and Virginia 1999).

Support for either of these ideas regarding the importance of diversity in soil is scarce. However, one study of special note is that of Laakso and Setälä (1999*b*). These authors used model ecosystems to manipulate trophic group structure (comparing treatments composed of fungal-feeding and microbi-detritivorous animals, and their combination), and species richness (one versus five species) within trophic groups, and examined effects on nutrient mineralization and plant growth. It was found that trophic group identity, and richness, and species richness within trophic groups, had no detectable effect on either nutrient mineralization or plant growth. The study also showed that compositional effects of soil faunal species within trophic groups were far less important than across trophic group effects, and that certain species, in particular the enchytraeid worm *Cognettia sphagnetorum*, had a functionally irreplaceable role in the decomposer system. Similar conclusions also arise from a modelling study by Hunt and Wall (2002). They used a simulation model based on short-grass prairie, that included 15 functional groups of microbes and soil fauna, and they found that while the 'whole community' was important for maintaining plant productivity in these models, the deletion of no single faunal group affected this measure significantly. Furthermore, deletion of only three functional groups affected N mineralization by up to 10%. This result was taken to suggest that ecosystems could sustain the loss of some functional groups with little decline in ecosystem function, due largely to compensatory responses in the abundance of surviving groups. Together, therefore, these studies support the notion that there is a high degree of functional redundancy in soil food webs at the species level—supporting the redundant species hypothesis—and also that some soil biological processes are driven by particular animal species that are functionally irreplaceable. Few studies have examined the role of microbial diversity as a driver of soil processes, although a study by Setälä and McLean (2004) tackled this issue by examining how the number of fungal taxa (using a gradient from 1 to 43 taxa) influenced decomposition of raw forest humus in microcosms. They found that decomposition, measured as CO_2 production, increased with increasing fungal diversity, but only in the species-poor end of the gradient, again suggesting a considerable level of functional redundancy among decomposer fungi.

Another microcosm study by Cragg and Bardgett (2001) came to similar conclusions, that is, differences in composition of the food web are more important than changes in diversity *per se*. These authors manipulated species richness and composition within a model soil community of up to three species of fungal-feeding Collembola. They showed that, when in monoculture, individual species differed markedly in their effect on plant litter decomposition and nutrient release, but the effects of two and three species combinations on these processes were due to differences in the composition of the collembolan community, and in particular the presence of one, fast-growing collembolan, rather than the number of species present (Cragg and Bardgett 2001). In another microcosm experiment, Cole et al. (2004) examined the effects of individual species of microarthropods, and variations in microarthropod diversity of up to eight species, on soil microbial properties and plant uptake of an added organic ^{15}N source (glycine). They detected effects of increasing species richness of microarthropods on microbial biomass and mycorrhizal colonization of plants, but no effects on plant ^{15}N uptake. However, these effects on microbial properties of soil could be interpreted in relation to the presence of individual species, rather than diversity *per se*. Together, the findings of these studies suggest that soil processes are mainly driven by the physiological attributes of the dominant animal species present. This supports the notion that the effects of declining species diversity within a trophic group on soil processes are likely to be idiosyncratic, depending on which species are removed from the community. Such findings concur with studies of above-ground communities which point to the role of vegetation composition and dominant plant species as a major driving force of ecosystem function at local scales (Grime 1997; Hooper and Vitousek 1997; Wardle et al. 1997a).

Other studies, however, do point to important effects of faunal diversity on soil processes. For example, Liiri et al. (2001) found that total N uptake and growth of birch trees increased asymptotically with increasing richness of microarthropod species, an effect that was presumably due to increased soil

N availability with increasing species richness. In the same experiment, however, increasing species richness did not modify the stability of the ecosystem, measured as the effect of drought on birch growth. Griffiths et al. (2000) also found some evidence to suggest that alteration of soil biological diversity can affect soil processes. These authors reduced total soil biodiversity using varying time periods of chloroform-fumigation, and found that certain measures, such as nitrification and methane oxidation, were reduced as a consequence. In contrast, other measures of biological function in soil increased with fumigation time (e.g. N mineralization), or were not affected (e.g. decomposition), suggesting that effects of fumigation and changes in soil biological diversity, are process specific. It was also shown in this study that the biological community of the fumigated soil was less resistant to a second stress (i.e. copper addition). In a related experiment, Griffiths et al. (2004) manipulated soil microbial community structure by re-inoculating sterile soils with different levels of dilution of a non-sterile soil suspension, resulting in a gradient of biodiversity. They then subjected these soils to heat and copper stress, and found that less biologically diverse soils (i.e. those receiving the most diluted inoculum) were the least resistant to these stresses. While these studies provide evidence for diversity effects, it has been argued that the methods of manipulation used make the data susceptible to covarying diversity factors that confound the interpretation of results (Griffiths et al. 2000; Mikola et al. 2002). In other words, it is difficult, if not impossible, to disentangle the effects of diversity from covarying factors on ecosystem processes; for example, chloroform-fumigation and dilution procedures selected for certain species and thus cannot separate species composition effects from species richness effects (Griffiths et al. 2000).

Heemsbergen et al. (2004) also detected soil biodiversity effects on decomposition processes in laboratory studies, but found that they were explained by the *functional dissimilarity* of component species, rather than by species number. In microcosms, these authors found no conclusive effects of varying species number of macro-detritivores on a range of decomposition processes, such as mass loss, respiration, leaf litter fragmentation, and nitrification. However, they detected a significant positive relationship between soil respiration and litter mass loss, against the functional dissimilarity of the animal community, defined as the degree to which component species are functionally different in those processes, determined from their performance in monocultures. These effects were attributed to *facilitative interactions* between species, which were greater when component species differed in their functional role. For example, facilitation was detected in communities that contained the earthworm *L. rubellus*, which is especially effective in transporting litter, which has been fragmented by isopods and millipedes, deep into soil where it is subject to microbial attack. In contrast, inhibition of decomposition occurred between the functionally similar *Oniscus asellus* (Isopoda) and *Polydensmus deticulatus* (Diplopoda), suggesting competition for leaf litter between these species. Overall, these findings suggest that it is the degree of functional differences between species that is a driver of ecosystem processes, rather than species number. Therefore, to predict the consequences of species loss in soil requires an understanding of how individual species contribute to multiple species interactions in the community (Heemsbergen et al. 2004).

Apart from the above study, there appears to be little support in the literature for a predictable relationship between soil biological diversity and process rates that are crucial to ecosystem function. Rather, most evidence suggests that changes in the abundance of particular species and alteration in the nature of multiple species interactions that occur in soil are the main biotic control of ecosystem function. There is some evidence for a high degree of redundancy within soil food webs—an argument that is bolstered by the prevalence of omnivory in soil food webs (Scheu and Falca 2000; Ponsard and Arditi 2000; Maraun et al. 2003)—but it is also clear that some species are more redundant than others in diverse soil communities.

The absence of consistent diversity effects in experiments is likely to be due to a number of reasons related to the way that they are performed. Most notably, experiments are typically done under very artificial and structurally simple conditions and use a limited range of organisms that vary greatly in their life history and their performance in microcosms. As already discussed, soil food webs in nature are highly complex and involve a multitude of interactions that cannot be observed under simple laboratory conditions. Also, in such a heterogeneous environment as the soil, effects of biodiversity on soil processes will vary greatly in space and time, as will the relative importance of biodiversity to abiotic factors in soil. A study by Liiri et al. (2002) stresses this point; in a three-year field lysimeter study located in a Boreal forest, abiotic factors, such as moisture and temperature, were considerably more important as determinants of soil processes than were variations in the structure and biomass of the faunal community in soil. Similarly, Gonzalez and Seastedt (2001), in a comparative study of controls on decomposition in different ecosystems, found that the role of soil fauna varied between ecosystems, being disproportionately greater in tropical wet forests than in tropical dry or subalpine forests. These findings indicate, therefore, that while diversity and community composition of soil organisms might be important regulators of ecosystem processes at the local scale, abiotic factors such as climate are likely to be stronger regulators at larger regional scales (Mikola et al. 2002). This is analogous to the situation for plant productivity, which is driven mainly by abiotic factors at a larger scale, but is also affected by species composition and species richness at a smaller scale (Huston 1994).

In making the above conclusions it is important to note that the studies that have examined relationships between soil biodiversity and ecosystem function have largely considered soil animals, notably detritivores. As a consequence, very little is known about the functional role of changes in microbial community composition and diversity. The general view is that redundancy within microbial communities varies for different functional groups of microbes; influences of microbial community composition and diversity are most likely to be observed for processes that are physiologically or phylogentically 'narrow', such as N-fixation or nitrification, whereas 'broad' processes such as N mineralization should be insensitive to microbial community composition (Schimel 1995). This view, however, has been challenged by Schimel et al. (2005). These authors argue that 'broad' processes of mineralization and immobilization can be 'aggregated' into individual components—based on specific enzyme activities and spatial distribution of microbes in microsites—that are sensitive to microbial community composition. In other words, different classes of enzymes involved

Box 3.4 Linking phenotypic data on microbes to natural processes

Disentangling the specific functions of individual microbes within complex microbial communities is not at all straightforward. However, our ability to make mechanistic linkages between phenotypic data on micro-organisms and natural processes has recently been advanced by the development of techniques that measure the assimilation of isotopically labelled substrates by biomarkers of the microbial community (Radajewski et al. 2000). The principle of these methods is that active microbial populations utilizing a particular substrate with a radio- or stable isotopic label, such as ^{14}C- or ^{13}C- methane, will incorporate the labelled isotope into cell biomarkers, such as phospholipids fatty acids (PLFAs), amino acids, or DNA/RNA. These isotopically labelled biomarkers can then be separated from unlabelled material using a range of techniques (e.g. gas chromatography coupled with isotope ratio mass spectrometry) to identify which microbial groups use these C sources to support their growth. For example, ^{13}C-labelling of PLFAs was used by Boschker et al. (1998) to link microbial populations to the process of methane oxidation in aquatic sediments, and the same approach was used by Bull et al. (2000) to identify micro-organisms involved in the oxidation of atmospheric levels of methane in a forest soil. Similarly, Radajewski et al. (2000) separated ^{13}C-DNA from bacterial cells that were exposed to ^{13}C-methanol, demonstrating two distinct groups of eubacteria that utilized methanol, and Ostle et al. (2003) estimated turnover of root derived ^{13}C in soil bacterial RNA. These methods are still being refined, but what they clearly offer is the possibility of identifying the active involvement of micro-organisms in specific processes *in situ*.

in these processes are produced by different groups of micro-organisms, and different micro-organisms may live and function in different types of microsites. These authors argue that as long as we view N mineralization as a simple 'aggregate' process, we will be blind to the specific roles of microbial community composition in regulating N cycling (Schimel et al. 2005). This emphasizes the importance of the need to disentangle the specific functions of microbes involved in important soil functions (Box 3.4).

3.6 Conclusions

Microbes have many important roles to play in the ecosystem processes of decomposition and nutrient mineralization, and these activities can determine the availability of nutrients to plants. However, a full understanding of the role of soil biota in driving these processes requires consideration of the interactions that occur between microbes and other organisms within the soil food web. There is ample evidence to prove that the feeding activities of microbial-feeding fauna can greatly influence the abundance and activities of microbes, ultimately affecting nutrient flux. Also, higher-level consumers can have cascading effects on soil food webs that ultimately affect microbes and the processes that they drive. Soil animals also modify greatly the physical structure of soil, thereby affecting the habitat and activities of other fauna and microbes, and also influencing the physical movement of water and solutes through soil. While variations in soil food web diversity have the potential to influence ecosystem properties, experimental evidence indicates that relationships are highly idiosyncratic, and that while there may be a high degree of redundancy within soil food webs, some species are much more important than others. There is emerging evidence, however, that biodiversity effects on soil processes can be predicted by the degree of functional differences among species. Much is still to be learnt about the functional role of soil food web interactions, especially of microbes, and also the relative importance of these biotic forces to abiotic ones as drivers of ecosystem function.

4 Linkages between plant and soil biological communities

4.1 Introduction

Soil organisms are intimately linked to the plant community. Not only do plants provide C and other nutrients to the decomposer community, but plant roots also act as a host for many soil organisms, such as herbivores, pathogens, and symbionts. The soil biota in turn influence plant communities indirectly by recycling dead plant material and making nutrients available for plant use, and directly through the action of the root-associated organisms which selectively influence the growth of plant species, thereby affecting plant productivity and community structure (Fig. 4.1). The increasing recognition of the influence of these components on each other has drawn the attention of ecologists to the role played by these above-ground–below-ground feedbacks in controlling ecosystem processes and properties (Wardle et al. 2004a).

This chapter examines the various ways that plants and soil biota influence each other, and highlights some of the recent findings on the links that exist between plants and soil biological communities. First, there is a discussion of some of the ways that plants, and changes in the diversity and composition of plant communities, can modify soil biological communities and their activities. Second, the issue of how soil biota can in turn influence plant growth and also act as agents of vegetation change, both in the short- and longer-term is considered. The ultimate goal of this chapter is to illustrate the nature and ecological significance of linkages between plant and soil communities at the community and ecosystem scale.

4.2 Individual plant effects on soil biological properties

The principal route through which plants modify soil food webs is through their influence on the quality and quantity of organic matter that is returned

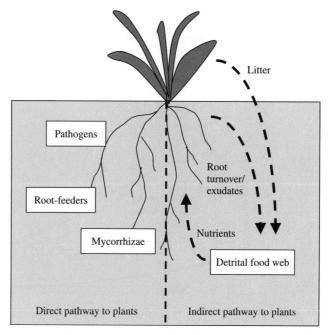

Fig. 4.1   Above-ground communities as affected by both direct and indirect consequences of activities of soil food web organisms. Right-hand half of figure: feeding activities in the detritus food web stimulate nutrient turnover, plant nutrient acquisition, and plant performance. Left-hand half of figure: soil biota exert direct effects on plants by feeding upon roots and forming antagonistic or mutualistic relationships with their host plants.

to soil, in the form of plant litter and root exudates. Plant species differ greatly in their basic ecological traits and this variation is reflected in differences in the quantity and quality of resources that they produce, in terms of both litter inputs (shoot and root), which impact on soil food webs over relatively long timescales and large spatial scales, and root exudates, which operate more in the short-term over finer spatial scales (i.e. the rhizosphere). Plant inputs to soil also vary greatly in terms of the relative proportions of different carbon constituents that they contain, which degrade at different rates reflecting their resource value to the decomposer organisms (Fig. 4.2). These can be broadly characterized into three carbon components of differing degradability: (1) the most easily degradable carbon is the labile, low molecular weight fraction that includes sugars and amino acids, which provide a soluble and readily usable resource for microbes; (2) the intermediate fraction includes moderately labile substances such as cellulose and hemicelluloses, which are the most abundant constituents of plant residues; and (3) the most recalcitrant fraction are structural materials, such as lignin. Therefore, the extent of plant species identity influences on soil biota and their activities will be maximized when differences in such ecological traits between species is greatest, since this will maximize differences in terms of the quantity and quality of plant-derived resources entering the soil.

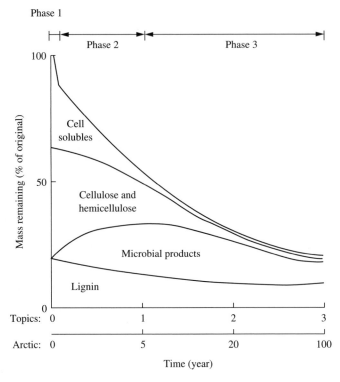

Fig. 4.2 Time course of leaf litter decomposition in relation to carbon components of differing degradability and the timescales commonly found in warm tropical and cold arctic conditions. Phase 1 is dominated by leaching of soluble compounds, whereas phase 2 involves degradation of labile, low molecular weight fractions, such as sugars and amino acids, and then of moderately labile substances such as cellulose and hemicelluloses, which are the most abundant constituents of plant residues. Finally, the most recalcitrant fraction, such as lignin, is degraded. (Redrawn with permission from Springer-Verlag; Chapin et al. 2002.)

4.2.1 Experimental evidence of effects of individual plants on soil biota

A number of laboratory experiments have shown that the growth of individual plant species in soil can influence the size of the microbial biomass and also lead to the selection of specific microbial communities in the root zone. For example, Bardgett et al. (1999c) performed a pot experiment that showed that single species of grassland plants differ markedly in their impact on soil microbes, with different compositions and abundances of microbial communities in their rhizospheres. Similarly, Innes et al. (2004) showed that plant species of grassland had markedly different effects on the abundance and structure of rhizosphere microbial communities, depending on the soil type that these plants were grown in; bacteria were positively affected by the growth of various grasses and herbs in fertile grassland soil, but the same plants negatively affected these microbes when grown in less fertile soil. Other authors have shown interspecific variability in the way that plants influence the size, activity, and structure of rhizosphere microflora (Wardle and Nicholson 1996; Grayston et al. 1998), and that populations of consumer organisms (e.g. nematodes) differ between plant species (Griffiths et al. 1992). Furthermore, these effects do not appear to be related to plant productivity, suggesting that they are attributable to species differences in the quantity and quality of exudates released from roots. That root exudates act as major drivers of such changes is supported by a laboratory study by Griffiths et al. (1999). These authors showed that variations in the composition and quantity of synthetic root exudates added to soil held at constant moisture greatly affected the structure of soil microbial communities.

Effects of individual plant species on microbes and their consumers have also been shown to occur under field conditions. In tallgrass prairies of North America, for example, sampling of soils under various C3 and C4 grass species revealed that some bacterial and nematode groups were responsive to the presence of particular plant species (Porasinski et al. 2003). Experiments in New Zealand pastures showed that the community structure of soil microbes, microbe-feeding nematodes, and herbivorous nematodes and arthropods, were all responsive to the removal of particular subsets of flora from the plant community (Wardle et al. 1999*a*). In a long-term grassland management experiment in northern England, Smith et al. (2003) showed that changes in microbial community structure in soil, and especially an increase in the relative abundance of fungi, were related in part to the introduction of particular plant species into the plant community. As in pot experiments, these effects are likely the result of differences in root turnover and root exudation patterns of grassland plants, and also due to variations in resource quality of litter inputs from roots and shoots.

Individual plants have also been shown to have large effects on soil biological communities in strongly nutrient limited situations. For example, in high arctic communities Coulson et al. (2003) observed that the local distribution pattern of soil microarthropod communities was strongly affected by the presence of particular plant species; soil microarthropod density was found to vary significantly in soils beneath different plant species, and distinct microarthropod assemblages were associated with particular species. In another study, Bardgett and Walker (2004) examined the effects of colonizer plant species on microbial community growth and composition on recently deglaciated terrain at Glacier Bay, south-east Alaska (Figs. 4.3 and 4.4). Analysis of microbial communities using phospholipids fatty acid analysis (PLFA) revealed that, of the colonizer species, *Alnus* and *Rhacomitrium* had the greatest impact on microbial growth, increasing total

Fig. 4.3   Photograph of the recently deglaciated terrain, ice-free for 12–15 years, on the northern shore of Upper Muir Inlet, at the head of the Muir Glacier, south-east Alaska. The plant in the foreground is *Dryas drommundii*, whereas the large shrub is *Alnus sinuata*. (Image by Richard Bardgett.)

microbial biomass by some six- to seven-fold relative to bare soil. These colonizer species also had significant effects on the composition of their associated microbial communities, dramatically increasing fungal relative to bacterial growth in soil. These findings were taken to suggest that certain colonizers—especially *Alnus* and *Rhacomitrium*—of recently deglaciated terrain are instrumental in initiating the development of microbial communities in soil, by promoting total microbial growth and facilitating the introduction of fungi into the soil microbial community. This early stage of microbial succession, which is promoted by these particular colonizers, may be important in accelerating ecosystem development and promoting efficient decomposition and recycling of nutrients.

4.2.2   Hemiparasitic plants

There is growing evidence to suggest that parasitic plants may also have important effects on soil biological properties. There are some 3000 known species of parasitic angiosperms that occur in most natural and semi-natural ecosystems from the high Arctic to tropical forests (Press and Graves 1995), and in these places they are known to have a range of negative effects on their hosts plants (Press 1998). This, in turn, has the potential to influence plant community structure and productivity, and hence soil biological properties. The potential of hemiparasitic plants to influence plant community attributes is evidenced from studies of semi-natural grasslands

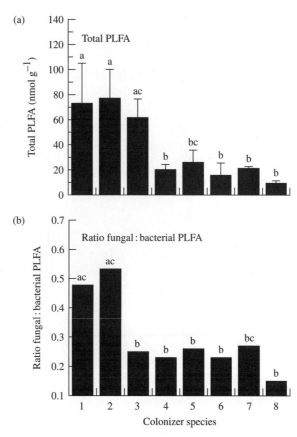

Fig. 4.4   Impact of colonizer species on microbial communities of recently deglaciated terrain at Glacier Bay, south-east Alaska: (a) Total PLFA, a measure of microbial biomass, and; (b) the ratio of fungal : bacterial PLFA, a measure of shifts in the relative abundance of fungi and bacteria within the whole microbial community. Values with the same letter are not significantly different from one another. (Data from Bardgett and Walker 2004.) Species: 1 = *Alnus sinuata*, 2 = *Rhacomitrium canescens*, 3 = *Equisetum variegatum*, 4 = *Dryas drummondii*, 5 = *Epilobium latifolium*, 6 = *Salix alaxensis*, 7 = *Salix arctica* 8 = *Bare soil*

in Britain and Europe. Here, the presence of the hemiparasitic plant *Rhinanthus minor* is correlated with local decreases in productivity and an associated increase in plant species richness (Pywell et al. 2004). This is because the hemiparasite modifies the competitive relationships between component species, reducing the dominance of fast-growing grasses, thereby increasing establishment and survival of slower-growing forbs. These effects in turn have the potential to influence soil biological properties via changes in the quantity and nature of resource inputs to soil.

A series of recent studies have revealed another mechanism by which parasitic plants can have positive effects on plant communities, notably via

Fig. 4.5   The common hemiparasitic plant *Bartsia alpina* of subarctic communities that increases total N inputs from litter to soil by some 42%. (Image by Helen Quested.)

enhanced nutrient soil cycling. In subarctic communities a range of hemiparasites have been shown to accumulate nutrients in their leaves, thus producing high quality plant litter that releases nutrients that would otherwise remain in the host plant or in slowly decomposing plant litter (Quested et al. 2002). Laboratory experiments showed that this leads to increased rates of decomposition of litter from non-parasitic plants (Quested et al. 2002) leading to enhanced nutrient availability in soil and plant growth (Quested et al. 2003*a*). Related field studies revealed that in these subarctic communities, the common hemiparasite *Bartsia alpina* (Fig. 4.5), increased total N inputs from litter to soil by some 42%, providing clear evidence of a novel mechanism by which these species influence soil properties in the ecosystems in which they occur (Quested et al. 2003*b*). Further studies are required to establish the effects of these parasitic plants on soil biota and their function in other ecosystems, and on local patterns of soil biodiversity within patches where these plants are present.

4.2.3   A role for plant polyphenols

Discussion of how plant species modify soil properties has in recent years pointed to a possible role for polyphenols, which are the most widely distributed class of plant secondary metabolites (Box 4.1). Traditionally, ecological research on plant-produced polyphenols has focused on their role as anti-herbivore compounds, defending plants against herbivore attack (Hartley and Jones 1997). Recently, however, it has been recognized that the production of these compounds by particular plant species could also have important effects on soil biota and nutrient availability, with far reaching consequences for the growth and competitive status of the plant that produced them (Hättenschwiler and Vitousek 2000). A number of studies have shown that polyphenols from specific plant species can directly influence certain groups of soil organisms. For example, Baldwin et al. (1983) showed that plant polyphenols can inhibit nitrifier bacteria and

Box 4.1   Polyphenols and their fate in soil

Phenolic compounds have many roles in plant biology, including UV protection, herbivore and pathogen defence, and contribution to plant colouration. They are defined by the presence of at least one aromatic ring bearing one (phenol) or more (polyphenols) hydroxyl substitutions. Polyphenols are often divided into two groups: (1) low molecular weight compounds, such as simple phenols, phenolic acids, and flavonoids; and (2) oligomers and polymers of relatively high molecular weight, such as tannins. Low molecular weight phenolics occur universally in higher plants, whereas higher molecular weight compounds are most abundant in woody plants, and often absent in herbaceous plants. Total polyphenol concentration in plants varies from 1 to 25% of total green leaf dry mass. They enter the soil via two pathways: (1) as leachates from stem, leaf, and root material; and (2) within leaf, stem, and root litter. While little is known on the relative importance of these routes, it is generally assumed that significantly more polyphenols enter soil via decomposing litter than through leachates. For example, in sugar maple (*Acer saccharum*) stands some 23 kg ha^{-1} an^{-1} of soluble polyphenols were leached from the canopy, whereas 196 kg ha^{-1} an^{-1} entered the soil via plant litter (MacClaugherty 1983). On entering the soil, soluble polyphenols are faced with a variety of fates: they may be degraded and mineralized by heterotrophic microbes; they can be transformed into insoluble and recalcitrant humic substances; they might be adsorbed to clay minerals or form chelates with Fe and Al ions; or they might remain dissolved, and be leached out of the soil and into waterways as dissolved organic material. Insoluble polyphenols enter the soil as litter, and this is usually the main form of polyphenol that enters soil. These are slowly decomposed by microbes, contributing to the soluble polyphenol fraction. Some fauna can digest polyphenols in their guts (e.g. earthworms), thereby enhancing their breakdown or mixing them with minerals to form organo-mineral complexes.

Leake and Read (1989) showed that mycelium biomass of an ericoid mycorrhiza was reduced by a mixture of common phenolic acids, but was increased by host-derived polyphenols from shoot extracts. All these studies on below-ground effects of polyphenols, however, were done under laboratory conditions, so the relevance of these effects at the ecosystem scale is uncertain.

There is a growing body of literature that suggests that species-specific production of polyphenols could have marked effects on soil properties at the ecosystem scale, ultimately affecting competitive interactions of plants in natural ecosystems. For example, Schimel et al. (1998) showed in Alaskan taiga forest stands that polyphenols produced by balsam poplar (*Populus balsamifera*) inhibited N fixation by alder (*Alnus tenuifolia*) and increased microbial immobilization of N, which was suggested as the cause for a decline in growth of alder with increasing dominance of poplar (Fig. 4.6). These findings suggest that plant polyphenols produced by poplar are a control on soil nutrient dynamics and species interactions in these forest systems. Another example is that of Northup et al. (1995) who studied the effects of polyphenols released from Bishop pine litter (*Pinus muricata*) on soil nutrient availability. These authors found that high levels of polyphenols released from pine litter inhibited N mineralization, but increased the release of DON from pine litter (Fig. 4.7). On the basis of this finding, these

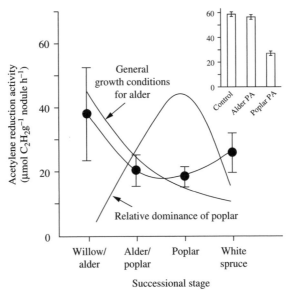

Fig. 4.6   Changes in N_2 fixing activity, measured as acetylene reduction activity of nodules, and general growth conditions of alder (*Alnus tenuifolia*) along a successional sequence of Alaskan taiga forest. N_2 fixing activity and growth of alder declined with increasing dominance of poplar, which was attributed to a negative effect of polyphenols (PA) produced by poplar on alder, as observed in the laboratory (see insert graph). (Redrawn from Hättenschwiler and Vitousek 2000, using data from Schimel et al. 1998)

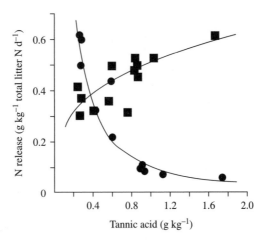

Fig. 4.7   Release of N as either inorganic N (●) or DON (■) from Bishop pine litter (*Pinus muricata*) as a function of the concentration of total phenolics in litter (tannic acid equivalents) during a three-week laboratory incubation study. Data show that high levels of polyphenols released from pine litter inhibit N mineralization, but increase the release of DON from pine litter. (Redrawn from Hättenschwiler and Vitousek 2000, using data from Northup et al. 1995)

authors suggested that plants living in N-limited ecosystems, such as Bishop pine, might actually benefit from increasing the DON : mineral N ratio in that it could lead to reduced N loss, owing to reduced leaching of inorganic N, and to a short-cutting of the microbial mineralization step by allowing increased uptake of organic N. As such, it was hypothesized that high polyphenol production by certain plants on infertile soils might represent an adaptive attribute to regulate the fate of N and to influence the plants competitive advantage for uptake of organic N.

There is also emerging evidence that plant phenolics have an important role in soil N cycling and plant competition in alpine ecosystems. Bowman et al. (2004) studied a potential soil feedback involving the slow-growing alpine herb *Acomastylis rossi*—that produces litter that is extremely rich in phenolics—and its co-dominant in fertile meadows, the fast-growing, N-demanding grass *Deschampsia caespitosa* (Fig. 4.8). In a microcosm study, they found that phenolic compounds in litter of *Acomastylis* enhanced microbial activity in soil—since they acted as a C source for soil microbes—leading to net N immobilization, which reduced soil N availability. They also found that *Acomastylis* litter reduced the growth of, and uptake of N by, *Deschampsia* relative to when it was grown with its own litter. Coupled with the finding that rates of N mineralization in the field were 10-fold lower in soils dominated by *Acomastylis* relative to those dominated by *Deschampsia* (Stelzer and Bowman 1998), these findings were taken to suggest that *Acomastylis* litter has the potential to influence soil N availability in a way that facilitates its persistence in alpine meadows. In other words, since

Fig. 4.8 Fertile, wet meadow site at Niwott Ridge Long Term Ecological Research (LTER) site, Colorado Rocky Mountains, USA. In these communities, a soil feedback operates involving the slow-growing alpine herb *Acomastylis rossi*—that produces litter that is extremely rich in phenolics—and its co-dominant in fertile meadows, the fast-growing, N-demanding grass *Deschampsia caespitosa*. (Image by William Bowman.)

Acomastylis can tolerate lower N availability due to its slow growth and conservative N use, it gains a competitive advantage over its faster-growing, N-demanding neighbour, *Deschampsia*, which is negatively affected by a reduction in resource supply. Overall, these findings—and those of other studies discussed above—indicate that plant-produced polyphenols can be important factors determining plant competitive interactions through their effect on soil microbial activity and nutrient cycling.

4.2.4 Theoretical framework for explaining plant species effects on soils

Attempts have been made to provide a theoretical framework for explaining how individual plant species affect soil food webs and decomposition processes, on the basis of plant ecophysiological traits (Wardle 2002, 2005). It is well established among plant ecologists that plant species adapted to particular habitat conditions have different sets of ecophysiological traits (e.g. Grime 1979; Chapin 1980), and it is becoming clear that these sets of traits can also determine soil biological properties, especially the rate of decomposition of litters produced by particular species (Grime et al. 1996; Cornelissen and Thompson 1997; Wardle et al. 1998; Cornelissen et al. 1999). For example, comparative studies have shown that leaf traits such as palatability of foliage and its decomposability are positively related, in that palatable plant species generally produce litter that is of a higher quality for decomposers than do unpalatable species (Grime et al. 1996). Similar comparative studies show that traits such as plant growth rate, size, and longevity (Wardle et al. 1998), leaf tissue strength (Cornelissen and Thompson 1997; Cornelissen et al. 1999), and nutrient use efficiency (Aerts 1997) are strong determinants of litter decomposition rate in soil (Box 4.2).

Box 4.2   Factors controlling decomposition rate of litter

A hierarchy of abiotic and biotic factors controls decomposition. At the highest level of the hierarchy, climate determines the rate of decomposition, in that it is the primary determinant of soil moisture and temperature, which together affect rates of main physical, chemical, and microbiological reactions that control decomposition. In general, decomposer organisms and their enzymes are more active in warm, moist conditions, such as in the tropics, whereas they are less active under excessively cold and/or wet conditions, such as found in northerly ecosystems where peat accumulates. At the second level in the hierarchy, the landscape and nature of parent material largely influences decomposition rates. For example, all else being equal, decomposition is faster in neutral than acidic soil, and an abundance of clay minerals tends to reduce rates of decomposition, due to increased water holding capacity and in some cases waterlogging of soil. The third level in the hierarchy is the quality and quantity of the organic matter produced, that depends on the nature and composition of plant communities. The release of energy and nutrients stored in dead organic matter depends strongly on the proportion of support tissues rich in lignin and lignocellulose, the proportion of secondary chemical compounds, such as tannins or polyphenols, and the ratio of nutrients to carbon. While the relative importance of these factors is not fully understood, there are certain measures of litter quality that are commonly used as predictors of decomposition. Of these, the C : N ratio is often used as an index of litter quality, in that litter of a lower C : N generally decomposes faster than that with a high ratio. In more recalcitrant litters, the lignin : N ratio is often a good predictor of decomposition rate. Other plant traits can also be powerful predictors of decomposition rate of individual plant species. For example, Wardle et al. (1998) conducted a comparative study of a range of plant species and showed that decomposition rate of leaf litter and plant stem were negatively related to plant mass, vegetative growth rate, and positively related to stem N content. This suggests that plant traits can be used as powerful predictors of decomposition rate in ecosystems, and can act as alternatives to the litter quality characteristics that are more commonly used.

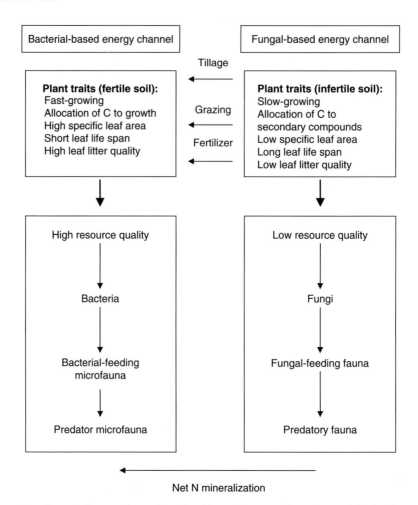

Fig. 4.9   The effects of plant species on fungal and bacterial energy channels in soil food webs, and their relation to plant traits and a range of other extrinsic factors.

The recognition of such relationships between sets of plant traits and decomposition processes has led to the suggestion that sets of traits of particular plant species are also likely to select for decomposer food webs with certain basic attributes (Wardle 2002, 2005). This idea has not been widely tested, but it recognizes that broad sets of traits linked to plant adaptation to resource availability have important consequences for litter quality, which in turn lead to dramatic effects on soil food webs. For example, it is generally accepted that fast-growing plant species that proliferate in fertile soils allocate most of their C to rapid growth, producing an abundance of foliage of high photosynthetic capacity, and producing easily decomposable litter that is rich in nutrients. In contrast, slow-growers that dominate conditions of low nutrient availability tend to be small, producing nutrient-poor foliage and litter that is rich in recalcitrant compounds such as lignin and phenolics. These contrasting plant traits have important implications for soil food-web structure and nutrient turnover, in that fast-growing plants are thought to select bacterial-dominated food webs associated with 'fast' cycling of nutrients, whereas slow-growers favour a fungal-dominated food web and 'slow' cycling of nutrients (*sensu* Coleman et al. 1983; Moore and Hunt 1988) (Fig. 4.9). The basis for this distinction is that high quality litter (low C : N) of fast-growing plants stimulates the growth and activity of bacteria and their consumers (e.g. bacterial-feeding protozoa and nematodes) leading to enhanced decomposition and nutrient turnover via the microbial loop (Chapter 3), which would further benefit rapidly growing plant species adapted for fertile sites. In contrast, poor quality litter (high C : N and high concentration of structural materials, such as lignin) produced by slow-growing plants favours the growth of fungi and their consumers (e.g. fungal-feeding collembolans and mites), which are typically associated with low rates of nutrient turnover, hence favouring still slower-growing plants that are adapted to infertile conditions. Overall, this framework suggests that there is a sense of mutualism between the plant and soil community, in that certain plant species can modify soil food webs and decomposition processes in a way that creates a more favourable habitat for that species, thereby increasing its competitive status.

4.3   Plant diversity as a driver of soil biological properties

In recent years, there has been much interest in the effects of plant species diversity and composition on ecosystem functioning. To date, most research in this area has focused on predicting the consequences of plant species richness on plant productivity, and the debates surrounding this issue are well rehearsed (see Loreau et al. 2002). There is now a growing body of literature that examines how plant species richness (living and dead plant parts) influences the structure and diversity of soil communities and their functioning. Theoretically, there are a number of ways through which plant diversity could influence soil biota and their activities (Fig. 4.10) (Wardle and Van der Putten 2002; Wardle 2005). Increasing plant richness could positively affect the diversity of soil biota through increased probability of including plants of different root morphological characteristics (e.g. different root densities and structures) as well as different root chemical properties (e.g. different litter types, root exudates, and tissue qualities and quantity). These differences may result in a greater diversity of food resources and habitat heterogeneity, creating available niches supporting diverse biotic assemblages enhancing both decomposition and nutrient mineralization.

While such theoretical arguments are compelling, the actual evidence for plant diversity effects on soil biological properties is very mixed.

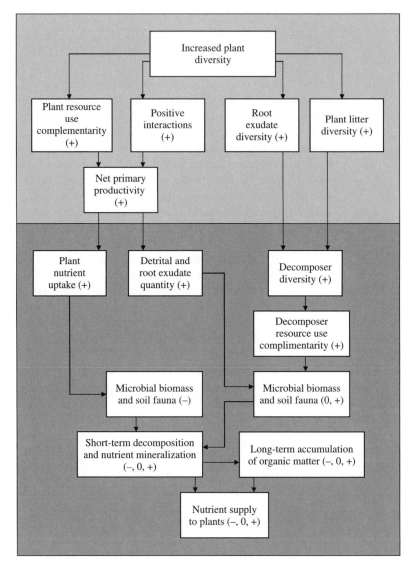

Fig. 4.10   Hypothetical mechanisms by which changes in plant species richness may affect decomposer-mediated processes. +, 0, and − indicate positive, neutral, and negative effects respectively. (Redrawn with permission from Oxford University Press; Wardle and Van der Putten 2002.)

For instance, it has long been known that decomposition rate varies due to differences in leaf litter quality between plant species and that decomposition rate changes when different leaf species are combined (e.g. Chapman et al. 1988; Blair et al. 1990). However, more recent studies that have manipulated species richness or the relative abundance of litters show no predictable effect on soil biota or decomposition (Wardle et al. 1997*b*; Bardgett and Shine 1999; Hector et al. 2000; King et al. 2002). A similar story comes from studies that have examined soil biological properties in experimental field trials where grassland plant species diversity and identity have been experimentally manipulated. In a study of an experimental grassland site in Germany, for example, Gastine et al. (2003) found no effects of plant diversity or functional group identity on a variety of soil biological properties, including soil N availability, microbial activity, and abundance and diversity of fauna. Similarly, Salamon et al. (2004) found in grassland field trials that neither plant species richness nor the number of plant functional groups affected total diversity of Collembola. However, they found that the number of Collembola increased in the presence of legumes, benefiting from high litter quality and increased microbial biomass in the rhizosphere of this plant. Wardle et al. (1999*a*) also found no predictable effect of manipulating grassland plant diversity—using *in situ* plant removal—on the diversity of microbes, microbe-feeding nematodes, predatory nematodes or soil herbivores, and De Deyn et al. (2004) showed in experimental grassland trials that individual plant traits that affect resource quality were more important than plant diversity as a driver of soil nematode community structure and diversity. These authors set up a three-year field experiment in which 8 plant species (4 forb and 4 grass species) were grown in monocultures and mixtures of 2, 4, and 8 plant species, along with communities with 16 plant species. They found that while both plant species diversity and plant identity affected nematode diversity, there was more variability in nematode diversity between monocultures of individual plant species than between different levels of plant species diversity, indicating that plant identity is more important than plant diversity. Zak et al. (2003*b*) sampled soils from a long-term (seven-year) field trial at Cedar Creek, USA, and found that microbial-community biomass, respiration, and fungal abundance were all affected by plant species diversity, increasing with greater plant diversity. However, these changes were related to higher levels of plant productivity in high diversity treatments, rather than diversity *per se*. It was also found that rates of N mineralization were affected by plant diversity, being greater in high diversity treatments, suggesting that enhanced N supply in these treatments contributed to increased plant production.

Studies done under controlled glasshouse conditions give overriding importance to plant species identity, rather than to species richness, as the primary driver of soil biota and processes. For example, Wardle et al. (2003*b*) conducted a glasshouse experiment in which three plant species of three functional groups (grasses, N-fixing legumes, and forbs) were grown in monoculture and in various mixtures. They found that plant species identity had important effects on biomass and population densities of primary consumers (microbes and herbivorous nematodes) and secondary consumers (microbe-feeding nematodes and enchytraeids), but the effects of plant species or functional group richness on these measures were few. In another study, Johnson et al. (2003*a*,*b*) set up a series of experimental treatments comprising bare soil, monocultures of a mycotrophic grass and non-mycotrophic sedge, and a species-rich mixture of four forbs, four grasses, and four sedges. Using a range of molecular tools, it was found that both AM fungal diversity (Johnson et al. 2003*a*) and bacterial diversity (Johnson et al. 2003*b*) were affected significantly by plant composition, suggesting that functional type of plants is a strong driver of microbial diversity and function in soil.

While most studies point to the important role of plant functional types as drivers of below-ground function, there is emerging evidence that above-ground and below-ground diversity operate interactively to determine ecosystem process rates. This was shown to be the case by Hättenschwiler and Gasser (2005) who tested how altering litter species diversity of coexisting temperate woodland plants affects species-specific decomposition rates and whether large litter-feeding soil animals control litter diversity—function relationships in microcosms. They found that decomposition of litter from a particular plant species changed greatly in the presence of litters from other coexisting species. However, their most important finding was that the presence of soil animals in microcosms determined the magnitude and direction of litter diversity effects. In general, the presence of millipedes had a significant effect on decomposition of slowly decomposing litters, whereas earthworms were more important for more rapidly decomposing litters. Hence, the relationship between species diversity in the litter pool and decomposition is highly dependant on the types of soil animals present, which, as argued Hättenschwiler and Gasser (2005), needs to be considered for the understanding of the functional significance of litter (and hence plant) diversity for ecosystem processes. In other words, the interplay between plant species diversity and soil animals are directly relevant for nutrient supply rates and ecosystem productivity.

In conclusion, the available evidence suggests that relationships between plant diversity and soil biological properties are unpredictable, and depend largely on the ecophysiological traits of individual, and often dominant, plant species that are present within the community, or on changes in net primary productivity. In other words, individual plant species and functional group characteristics that influence soil resource quality and other soil conditions seem to play a more significant role than plant species richness in structuring soil biota and their functioning in ecosystems. However, as argued by Hättenschwiler and Gasser (2005), the interplay between above-ground and below-ground diversity is also of key significance for ecosystem function, and hence needs to be considered for understanding the importance of diversity in terrestrial ecosystems.

4.4 Influence of soil biota on plant community dynamics

Understanding the factors that drive change in the species composition of natural plant communities has long been a major theme of community ecology. A variety of abiotic factors, such as climate, disturbance, and competition for light and nutrients (Tilman 1982), and biotic interactions such as those between herbivores and plants (Bardgett and Wardle 2003), have been identified as being major drivers of vegetation change. In recent years, however, there have been a growing number of studies that point to the importance of soil biota and their interactions with plants as major structuring forces in plant communities (Wardle et al. 2004*a*; Van der Putten 2005). There are three major types of biotic interactions between plants and soil biota that can influence plant community dynamics: (1) interactions between dead plant materials and the decomposer subsystem, leading to changes in soil nutrient availability; (2) effects of biotrophic interactions between living plant roots and their herbivores, pathogens, and symbiotic mutualists; and (3) microbial mediated niche partitioning of nutrient use by plants. How decomposer organisms act on plant growth by influencing the availability of nutrients in soil has been discussed in Chapter 3. This section focuses on the last two processes as drivers of vegetation dynamics, namely the role of biotrophic interactions and microbial mediation of resource partitioning.

4.4.1 Mycorrhizal associations and plant community dynamics

There is abundant evidence for mycorrhizal fungi benefiting the performance of their host plants. This is due to improved plant uptake of water and P (Smith and Read 1997) and also increased protection against attack by pathogens (Newsham et al. 1995) and above-ground herbivorous insects (Gange and West 1994), leading to increased fecundity (Newsham et al. 1994). However, while mycorrhizal fungi associate with the majority of plants within any community, they confer different degrees of benefit on certain plant species, thereby directly influencing the structure of plant communities. One of the first studies to illustrate this experimentally was by Grime et al. (1987) who assembled diverse grassland communities in microcosms and allowed them to develop in the presence and absence of AM fungi. The presence of mycorrhizal fungi led to a shift in plant community composition, reducing the dominance of *Festuca ovina* in favour of several subordinate herb species that benefited most from AM fungal infection. The net effect of this was a significant increase in plant species diversity due to a relaxation of plant competitive interactions.

Positive effects of AM fungi on subordinate plant species have also been shown to occur in the field. For example, Gange et al. (1993) applied a selective biocide to an early successional grassland community in southern England, which led to a suppression of AM fungal infection and an associated increase in the abundance of competitively dominant grasses at the expense of subordinate herb species, thereby reducing plant species richness. Several mechanisms have been proposed to explain increases in plant diversity resulting from AM fungal infection. Allen (1991) suggested that mycorrhizal fungi might increase plant diversity due to spatial heterogeneity of fungal infectivity in field soil, allowing mycotrophic (i.e. species that depend on mycorrhizal associations) and non-mycotrophic species to coexist in patches of high and low inoculum. Alternatively, Grime et al. (1987) suggested that AM fungi increase plant diversity due to interplant transfers of carbon and nutrients via hyphal links, which lead to more even distribution of resources within the plant community, reducing the ability of certain species to monopolize resources. The net effect of such nutrient distribution would be a more equitable competitive interaction between plant species, promoting species coexistence and greater plant diversity. This view is also consistent with findings of Van der Heijden (2004) that AM fungi promoted seedling establishment in perennial grassland by integrating emerging seedlings into extensive hyphal networks and by supplying nutrients, especially P, to the seedlings. AM fungi therefore act as a symbiotic support system that promotes seedling establishment and reduces recruitment limitation in grassland (Van der Heijden 2004).

Not all studies show that AM fungi promote plant diversity. Hartnett and Wilson (1999), for example, found the opposite to be the case in tallgrass prairie, where suppression of AM fungal infection using the fungicide Benomyl led to a large increase in plant species diversity. This was due to a reduced abundance of dominant, obligately mycotrophic C4 tall grasses and consequent competitive release of subordinate facultative mycotrophs. This suggests that active AM fungal associations reduce floristic diversity in tallgrass prairie due to the promotion of competitive dominance of mycotrophic species. Similarly, it has been proposed that in tropical rainforests ectomycorrhizal associations encourage dominance of certain tree species, at the expense of arbuscular mycorrhizal trees that are less able to acquire nutrients and tolerate pathogen attack, thereby reducing species coexistence (Connell and Lowman 1989). These studies suggest another mechanism for mycorrhizal effects on plant communities, whereby differences in host plant responses to colonization by mycorrhizal fungi result in changes in plant diversity if the dominant competitors are more strongly or weakly mycotrophic than their neighbours.

In all the above studies, effects of mycorrhizal fungi on plant diversity have been attributed to selective and beneficial effects on certain species, either relaxing or enhancing competitive dominance. However, recent evidence indicates that the costs and benefits of maintaining symbiosis with AM fungi vary greatly between plant species, in that individual plant responses can range from positive (mutualism) to neutral (commensalism), and even to negative (parasitism), suggesting that the symbiosis is better defined as a continuum from parasitism to mutualism (Johnson et al. 1997; Klironomos 2003). This suggests that in addition to differences in mycorrhizal dependency (Harnett and Wilson 1999), interspecies differences in the direction and intensity of response to mycorrhizal infection, along the mutualism–parasitism continuum, also act as structuring forces in plant communities. This idea has been tested in grassland communities (Klironomos 2003), but has not yet been examined in other types of ecosystem.

Variations in mycorrhizal fungal diversity may also strongly affect plant diversity. It is well known that mycorrhizal fungi are highly diverse (e.g. Daniell et al. 2001) and different species have been shown to induce different growth responses in plants (Van der Heijden et al. 1998*a*). Therefore, it would be reasonable to assume that variations in the diversity of mycorrhizal fungi would affect the diversity of the plant community. This idea was tested by Van der Heijden et al. (1998*b*) by manipulating the number of native AM species (1–14 species) in experimental grassland units containing 15 plant species. They found that plant diversity and productivity were positively related to AM fungal diversity (Fig. 4.11(a),(b)). The mechanism for this was thought to be related to a reported increase in AM hyphal length in the more diverse treatments, enabling more efficient exploitation and partitioning of soil P reserves, thereby relaxing plant competition and increasing plant productivity (Fig. 4.11(c)). It has been proposed by Van der Putten (2005), however, that these effects may only be observed in relatively nutrient-poor environments. In fertile soils, effects of AM fungal diversity may be neutral or even negative, when related to nutrient uptake only. This has been shown for ectomycorrhizal fungi by Jonsson et al. (2001). These authors found that in a low fertility substrate, ectomycorrhizal diversity enhanced biomass production of *Betula pendula*, but no effects were found in fertile substrate, while for *Pinus sylvestris* the effects of ectomycorrhizal species richness were negative even in highly infertile substrate. In a similar way, ericoid mycorrhizal fungi might even reduce plant species diversity in nutrient-poor soils due to monopolisation of the available nutrients by their host species (Northup et al. 1995).

4.4.2 Nitrogen-fixing organisms and plant community dynamics

Vascular plants with N-fixing symbionts are extremely widespread as early colonists in primary succession (Fig. 4.12), especially on glacial forelands where they are believed to be major facilitators of plant succession since they contribute most of the N to developing communities, raising soil N to levels needed to support later successional species (Walker and Del Moral 2003). The mechanism by which N-fixers increase the availability of soil N to other species has been attributed to both the build-up and decomposition

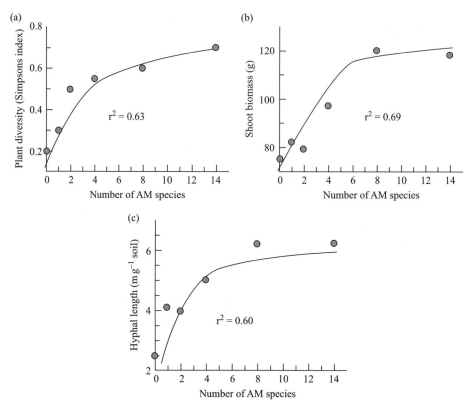

Fig. 4.11   Effects of manipulating the diversity (1–14 species) of native arbuscular mycorrhizal fungi on (a) plant diversity (Simpsons index), (b) shoot biomass, and (c) total fungal hyphal length in experimental grassland units. (Data from Van der Heijden et al. 1998*b*.)

of litter of high N content (Chapin et al. 1994; Kohls et al. 2003; Walker and Del Moral 2003). One of the classic examples of facilitation via N-fixers comes from the studies of primary succession at Glacier Bay, south-east Alaska (see Box 1.2). Here, Crocker and Major (1955) showed that *Alnus* shrubs increased soil N and soil organic matter content, and decreased soil pH, creating a more favourable environment for *Picea*, the next successional dominant. This view has been questioned since subsequent studies have revealed that other factors contribute to the replacement of *Alnus* by *Picea*, in particular competitive interactions between these species (Chapin et al. 1994).

There are many more examples of where N-fixers facilitate succession. For example in mined lands, trees perform better in soils where there have been N-fixers present (Walker and Del Moral 2003). In New Zealand sand dunes, *Lupinus* herbs have been shown to facilitate the growth of *Pinus* trees (Gadgill 1971) and in The Netherlands plant biomass on sand dunes increased following a stage dominated by the N-fixing shrub *Hippophaë*

(a)   (b)   (c)

Fig. 4.12   Vascular plants with N-fixing symbionts are extremely widespread as early colonists in primary succession. This slide shows dominant N-fixing plants of different primary seres: (a) *Alnus sinuata*, which forms thickets on deglaciated terrain at Glacier Bay, south-east Alaska; (b) *Coriaria arborea* colonizing recently deglaciated terrain at the Fox Glacier, South Island, New Zealand; and (c) *Coriaria arborea* colonizing floodplains near Kaikoura, South Island, New Zealand. (Images by Richard Bardgett.)

(Olff et al. 1993). Habitat and soil amelioration by N-fixers have also been demonstrated in primary succession on volcanoes, floodplains, and other glacial moraines (Walker and Del Moral 2003). It is also well documented that the role of N-fixers in succession is affected by other mutualists, namely AM fungi that can transport N from legumes through soil to neighbours via hyphal networks. For example, N transfer between agricultural legumes and grasses has been shown to occur via hyphal networks (e.g. Haystead et al. 1988) and it has been proposed—but not tested—that this route is of greater significance in N-limited pioneer plant communities where legumes are often strongly mycorrhizal (Haystead et al. 1988).

4.4.3   Root pathogens and plant community dynamics

One of the most widely known principles of agricultural management is that of crop rotation which prevents the build up of crop-specific pathogens in soil, which otherwise would lead to reduced yields of host plants. However, much less is known about the effects of soil borne pathogens on vegetation dynamics in natural ecosystems, although there is growing evidence to suggest that they can act as major agents of vegetation change through a negative feedback on their host plant. Evidence for this comes from a variety of sources. For example, in sand dune systems in The Netherlands, Van der Putten et al. (1993) showed that a build up of both root pathogens and root-feeding nematodes in the root zone of marram grass (*Ammophila arenaria*) led to the degeneration of this plant and its replacement by another species, namely *Festuca rubra*, that was not susceptible to these pathogens (Fig. 4.13). Similarly, seedling mortality of the invasive temperate tree *Prunus serotina* in tropical forests has been shown to be due to the build up

(a)

(b)

Fig. 4.13   In sand dune systems in The Netherlands, Van der Putten et al. (1993) showed that a build-up of both root pathogens and root-feeding nematodes in the root zone of marram grass (*Ammophila arenaria*) led to the degeneration of this plant and its replacement by another species, namely *Festuca rubra*, that was not susceptible to these pathogens. The photograph shows marram grass in its vigorous stage of growth (a) and in its degraded form, being replaced by *Festuca rubra* (b). (Images by Wim Van der Putten)

of pathogens of the genus *Pythium* (Packer and Clay 2000) and more recently, Klironomos (2002) showed that the rate of pathogen accumulation in the soil could determine the abundance and invasibility of plant species in grassland communities; rare plants tend to accumulate pathogens that limit their growth, whereas abundant and invasive species accumulate pathogens more slowly. Reinhart et al. (2003) also showed that invasibility of black cherry trees (*P. serotina*) into north-western Europe is facilitated by the soil community. In the native range in the United States, the build up of soil pathogens near black cherry inhibits the establishment of neighbouring conspecifics and reduces seedling performance of this tree, whereas in the non-native range, black cherry readily establishes in close proximity to conspecifics, and the soil community enhances the growth of its seedlings.

There is also much evidence of negative feedback playing an important role in grassland communities. For example, in glasshouse experiments, Bever (1994) and Bever et al. (1997) found that conditioning of soil by 9 out of 14 grassland plants led to negative feedback on that species, in that they performed worse in their 'own' soil than in that conditioned by another species. A number of mechanisms have been proposed to explain this negative feedback, including accumulation of host specific fungal pathogens (Mills and Bever 1998; Westover and Bever 2001), host specific shifts in the composition of rhizosphere bacteria (Westover and Bever 2001), and host specific changes in the composition of the AM fungal community (Bever 2002). It is highly likely that such negative feedback will have important consequences for plant diversity, in that it will prevent any single plant species from being dominant, thereby enabling more species to coexist within the plant community (Bever 2003; Reynolds et al. 2003). Such negative feedback mechanisms are thought to be of most significance in late successional communities, where plant host densities are greatest enabling sufficient build up of pathogens in soil (Reynolds et al. 2003).

4.4.4 Root-feeding fauna and plant community dynamics

As discussed in Chapter 2, there is a wide range of root-feeding fauna in soil, including nematodes, insects, and mites. As with symbionts, these organisms have differential effects on different plant species within communities, largely due to differences in palatability of root material. As a result, root-feeding fauna have the potential to greatly influence the dynamics of vegetation. This has already been discussed in relation to how root herbivory by plant-feeding nematodes and pathogens induced the degeneration of marram grass (*A. arenaria*), leading to its replacement by the grass *F. rubra* (Van der Putten et al. 1993). However, root-feeding nematodes have also been shown to influence small-scale shifts in vegetation composition of grassland (Olff et al. 2000), and to accelerate succession of grass species in agricultural grassland, especially when N becomes limiting after fertilization is stopped and plants become more susceptible to nematode attack (Verschoor 2001). Similarly, insect root herbivores have been shown to promote succession by reducing the success of early successional herb species in early successional grassland communities (Brown and Gange 1992), facilitating colonization by late successional species (Schädler et al. 2004). In model grassland communities, De Deyn et al. (2003) also showed that soil fauna have the potential to alter rates of succession in grassland communities. They showed that invertebrate root-feeding fauna enhanced secondary succession and local diversity in grassland communities by reducing the biomass of dominant species, enabling subordinate plant species to proliferate. Bradford et al. (2002) also showed that variations in the complexity of soil animal communities influenced the development of model grassland communities. These authors manipulated the complexity of soil faunal communities on the basis of body-size, with three treatments of increasing faunal complexity: (1) microfauna only; (2) microfauna and mesofauna; and (3) microfauna, mesofauna, and macrofauna. They found that addition of macrofauna (treatment 3) markedly affected plant community composition, significantly decreasing the biomass of forbs and legumes; in contrast, in the other two treatments, the biomass of forbs and legumes increased as the plant community developed. The precise mechanism driving these changes is not known, but it is most likely due to preferential feeding on N-rich legumes and forbs by large root-feeding fauna that are absent from the other treatments.

An additional role for root herbivores in plant succession concerns their effects on N transfers from early successional N-fixing plants. It has been reported that nematode feeding on a legume increased the transfer of N from this plant to a neighbouring grass species, and that the transfer was facilitated by increased microbial activity in the rhizosphere of the infested plant (Bardgett et al. 1999a). If these responses are common in legume–herbivore associations, then it is likely that they would contribute to facilitative effects of N-fixers in plant succession. For example, root-herbivore induced enhancements in nutrient flux in the rhizosphere of naturally occurring N_2-fixers, such as alder (*Alnus* species), may benefit other successional plant species, thus contributing to species replacement (Fig. 4.14). These ideas remain untested, but the widespread occurrence of N-fixers in primary succession and their strong association with host-specific root herbivores suggests that such interactions might have an important role in determining vegetation change and soil nutrient cycling in natural ecosystems.

Effects of root herbivores on plant communities may also manifest themselves through their influence on above-ground consumers of plant material and *vice versa*. Root herbivory has been shown to have positive (Gange and Brown 1989; Masters and Brown 1992; Masters et al. 1993, 2001), negative (Bezemer et al. 2004) and neutral (Moran and Whitman 1990; Salt et al. 1996) effects on the performance of foliar-feeding insects. Also, in the presence of foliar-feeders, the performance of below-ground herbivores has

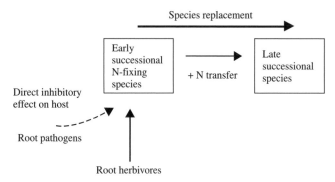

Fig. 4.14   Proposed facilitative role for root herbivores in plant succession through their positive effect on N transfers from early successional N-fixing plants to late successional plant species, contributing to plant species replacement. These facilitative effects are likely to occur in association with other, inhibitory, effects of root herbivores and pathogen attack on the early successional N-fixer, thereby contributing to the displacement of early successional species. (Redrawn with permission from Ecological Society of America; Bardgett and Wardle 2003.)

been found to be either reduced (Moran and Whitham 1990; Masters and Brown 1992; Salt et al. 1996) or remain unaffected (Masters et al. 1993). A wide range of mechanisms are likely to be responsible for these effects of above-ground and below-ground herbivores on one another. One plausible explanation is that root- and shoot-feeding insects indirectly influence each other via altering host plant allocation of primary metabolites such as N and carbohydrates (Gange and Brown 1989; Master et al. 1993). For example, root herbivory can induce a stress response in the host plant that can lead to the accumulation of amino acids and carbohydrates in foliage (Chapin 1991), thereby enhancing the performance of foliar-feeders. In turn, foliar-feeding can influence plant photosynthesis and C allocation to roots, thereby influencing the performance of root-feeders (Masters et al. 1993). There is also evidence that the concentration of foliar secondary plant compounds can increase following root damage and *vice versa* (reviewed by Van Dam et al. 2003). For example, infestation of roots of tobacco plants by the nematode *Meloidogyne incognita* has been shown to increase nicotine contents of foliage (Hanounik and Osborner 1977) and leaf damage can lead to increased nicotine in tobacco roots (Van Dam et al. 2001). More recently, Bezemer et al. (2004) showed that root herbivores of cotton plants reduced the performance of their above-ground counterparts due to the production of terpenoids following root herbivore attack. These studies demonstrate the multi-trophic nature of interactions between above-ground and below-ground organisms, and also the potential for root herbivores to influence their foliar counterparts and their antagonists. However, field studies are still needed to establish the significance of such interactions under natural conditions and also their importance for plant community dynamics.

4.4.5 Macrofauna and plant community dynamics

The vast majority of studies on effects of macrofauna on plant growth focus on the beneficial effects of earthworms on the growth of single plants, usually from an agricultural perspective (see Scheu 2003). (The many ways that large animals, such as earthworms, can positively affect soil fertility and plant growth are discussed in detail in Chapter 3.) While it has been recognized for some time that earthworms can have differential effects on plant functional groups, the consequences of this for vegetation development have hardly been investigated. The studies that have been done have looked at effects of earthworms on grassland plant communities, and generally show that the presence of earthworms favours the growth of fast-growing grass species over forbs and legumes. For example, Hopp and Slater (1948) documented that earthworms increased the dominance of grasses relative to legumes in experimental grassland systems. Similarly, Hoogerkamp et al. (1983) showed that the introduction of earthworms into polder soils, where they were previously absent, increased the performance of grasses relative to forbs, and Scheu (2003) presented data showing that the grass perennial ryegrass (*Lolium perenne*) benefited more from the presence of earthworms than did white clover (*Trifolium repens*) in pot experiments. The mechanism for this is not known, but it is likely to relate to a positive effect of earthworms on soil N availability that in turn promotes the growth of grasses over forbs and legumes. It has also been proposed by Scheu (2003) that grasses may benefit from a more patchy distribution of organic matter in soils with earthworms, since grasses tend to be better able to forage resource patches through root proliferation than are forbs and legumes (Hodge et al. 2000; Wurst et al. 2003).

Earthworms may also influence plant community dynamics by affecting seed transport and germination. Earthworms are known to consume large numbers of seeds, which are often deposited in a viable state in casts on the soil surface (Piearce et al. 1994). These casts can provide ideal conditions for seedling establishment of some plant species, being nutrient replete with reduced competition for light. In addition, germination of seeds can be promoted by passage through the earthworm gut, which often breaks dormancy (Scheu 2003). Therefore, seeds of plant species that are preferentially ingested and then deposited in casts may be at an advantage over other species in some plant communities, enabling them to better establish in mixed plant communities (Piearce et al. 1994). There is some evidence that earthworm casts can influence seedling recruitment in grasslands, especially of annual weeds. For example, the appearance of annual weeds was spatially linked to the position of casts in temperate grassland (McRill 1974), and experiments that involved the addition and/or removal of casts showed that they act as important seedbeds for weeds in grassland (Grant 1983). Virtually nothing is known about the significance of these processes in natural plant communities, but it is likely that seed transport and modification of germination by earthworms will substantially influence seedling recruitment in natural situations.

Other macrofauna can also have significant effects on the structure of plant communities, largely by creating patches of distinct vegetation. This is perhaps most noticeable in 'termite savanna', which is a term coined for vegetation formations in Africa and South America, consisting of grasslands and discrete islands of woodland based on large termite mounds (Harris 1964). In these ecosystems, termites can greatly influence vegetation succession. When they become stabilized, either because the colony has died out or because the mound is so large that only part of it is colonized, a vegetation succession occurs: grasses are first to colonize, followed by shrubs, and then, tall trees. When favourable conditions permit, these islands of woodland increase in size until they merge with each other and produce a closed canopy forest (Harris 1964). Ants similarly influence the patchiness of vegetation and succession through the formation of anthills; these structures have distinct vegetation from that which surrounds them and also their vegetation alters substantially when they are abandoned (Woodell and King 1991).

4.4.6 Microbial–plant partitioning of nutrients

N limits primary productivity in many ecosystems, and under such situations it is generally assumed that plants and microbes compete for soil N (Kaye and Hart 1997). New evidence is emerging to suggest that in N-limited situations plants and microbes may actually partition the N pool to avoid competition, both in terms of seasonal partitioning of N between the plant and microbial biomass, and also partitioning based on the chemical form of N. Evidence for seasonal partitioning of N comes from studies of alpine plant communities that are N-limited and show strong seasonality in plant growth. In the Colorado alpine tundra, for example, plants acquire N early in the season when they are actively growing and microbial N pools are low, whereas microbes immobilize N maximally later in the season after plant senescence (Jaeger et al. 1999). This general pattern has also been shown to occur on mountain sites of strong alpine-arctic character in the Highlands of Scotland (Bardgett et al. 2002), but the dominant plant *Carex bigelowii* meets its plant N requirements in early season through the use of internal reserves of N in roots/rhizomes, thus avoiding competition with microbes for soil DON (Fig. 4.15). In this ecosystem, significant net mineralization of N does not occur until microbial demands for N have been satisfied in June, corresponding to maximal plant biomass and shoot N content of *Carex*. Then, microbial sequestration of N is most intense in late season when plant demands for N subside after senescence. Little is known about what happens over the winter months. However, studies show that soil temperatures under snow pack remain sufficiently high to enable proliferation

Fig. 4.15   In high mountain communities, such as on high mountain plateau in the Cairngorms, Scotland, there is strong seasonal partitioning of N between plant and microbial pools. In this ecosystem, the dominant plant *Carex bigelowii* meets its plant N requirements in early season by using internal reserves of N in roots/rhizomes, thus avoiding competition with microbes for soil DON. Significant net mineralization of N does not occur until microbial demands for N have been satisfied in June, corresponding to maximal plant biomass and shoot N content of *Carex*. Then, microbial sequestration of N is most intense in late season when plant demands for N subside after senescence. (Image by Richard Bardgett.)

of large and diverse microbial communities, and that N immobilized by these microbes is released back into soil during and after snowmelt, thus becoming available to plants and summer microbial communities (Schmidt and Lipson 2004). The enormous size of this winter microbial community and its potential to sequester nutrients was revealed by Schadt et al. (2003). These authors sampled high altitude soils in the Colorado Rockies and were surprised to find that soil microbial biomass reached its annual peak in winter under snow, when it was dominated by a highly diverse fungal community.

Soil nutrients exist in a variety of forms in soil, both inorganic and organic. Evidence is emerging to indicate that in strongly N-limited ecosystems, such as arctic and alpine tundra, coexisting plants have the ability to partition a limited soil N pool, thereby reducing competition for this resource,

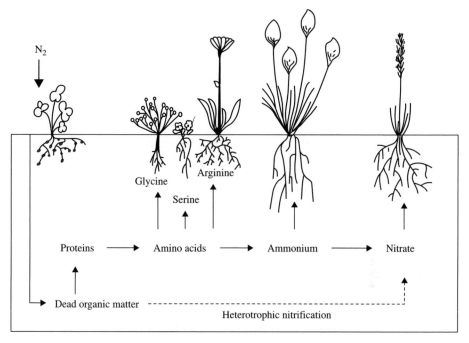

Fig. 4.16 Individual plant species access different chemical forms of N enabling species to coexist in terrestrial ecosystems.

through the uptake of different chemical forms of soil N, organic and inorganic (Fig. 4.16). This is especially significant since it suggests a mechanism for plant species coexistence, and hence the maintenance of botanical diversity in these ecosystems. Three lines of evidence support this notion. First, it is becoming increasingly evident that plants that inhabit N-limited ecosystems have the capacity to take up a diversity of chemical forms of N, both organic and inorganic (e.g. Schimel and Chapin 1996; Nordin et al. 2001; Henry and Jefferies 2003; Weigelt et al. 2003, 2005). Second, laboratory studies reveal that plant species of arctic and alpine tundra differ in their ability to take up different chemical forms of N, suggesting that species have fundamental niches based on N form (Miller and Bowman 2002, 2003). Third, coexisting species of arctic tundra have been shown to be differentiated in uptake of chemical forms of N, suggesting the existence of species' realized niches based on N form (McKane et al. 2002). The ability for plant communities to partition N on the basis of chemical form also requires that they can also compete effectively with soil microbes for these N forms (Bardgett et al. 2003), or alternatively that plants and soil microbes use different forms of N in soil, thereby avoiding competition. To date, these issues have not been fully tested.

The above studies examined niche partitioning of N, but a similar model could be envisaged for P, which similarly occurs in many forms in soil, both organic and inorganic (Turner et al. 2004b). As discussed in Chapter 3, plants access P via mycorrhizal fungi and through microbial enzymes. If microbes or their enzymatic capabilities are differentially associated with plant species, there is the potential for a type of resource partitioning to occur whereby plant species access different pools, thus avoiding competition for the same pool. The ability of plants to use different pools of P may also be facilitated through mycorrhizal fungi, which are known to be very diverse and to have a degree of host specificity (see Section 4.4.1). Such resource partitioning has been implicated, but not tested, as a means by which AM fungi promote plant diversity in temperate grassland, but further tests are needed to establish its significance in terrestrial ecosystems.

4.5   Plant–soil feedbacks and ecosystem development

So far, a range of examples have been given that show how soil biota and processes may influence vegetation dynamics in mainly early successional communities. In the longer run, in the prolonged absence of major disturbances, ecosystems typically enter a decline or retrogressive phase, during which there is a significant reduction in ecosystem productivity and standing plant biomass. As discussed in Chapter 1, these declines in primary productivity have been linked to long-term changes in the availability of soil nutrients that occur after prolonged soil development, especially the leaching and occlusion of P in strongly weathered soils, leading to P limitation of plant growth (Fig. 1.10) (Walker and Syers 1976; Crews et al. 1995). Declines in primary productivity have also been attributed to reductions in litter nutrient content and the build-up of plant phenolic compounds in the litter (Crews et al. 1995), that together limit the activity of decomposer organisms and the mineralization of nutrients, leading to a negative feedback on plant growth.

This idea was tested by Wardle et al. (2004b) across six long-term chronosequences formed by very different agents of disturbance, in Australia, Sweden, Alaska, Hawaii, and New Zealand. Each of these sequences had previously been shown to extend for sufficient duration for a decline in standing plant biomass to occur, and ranged in length from 6000 to over 4 million years, representing a range of locations, macroclimatic conditions, parent materials, and agents of disturbance forming the chronosequence. All six sequences were found to show a distinct decline phase (which usually becomes apparent a few thousand years after chronosequence initiation), associated with declines in tree basal area and increasing humus N : P ratios, pointing to increasing P limitation relative to N over time;

at three of the sequences, litter N : P ratios of dominant tree species also increased, again suggesting substrate P limitation for decomposers. This is consistent with N being biologically renewable while P is not (Chapter 3), and with P often being subjected to leaching and occlusion in strongly weathered soils (Crews et al. 1995). As a reflection of this, these changes in substrate quality were associated with significant reductions in the biomass of decomposer microbes, resulting in reduced rates of litter decomposition. These findings together suggest that as ecosystems develop—in the absence of major disturbances—P becomes increasingly limiting, setting a negative feedback in motion whereby low foliar and litter nutrient status reduces decomposer activity, which further intensifies nutrient limitation, thereby leading to ecosystem decline (Wardle et al. 2004*b*). The idea that microbial processes of decomposition are limited by P availability in forest ecosystems is also supported by studies of Cleveland et al. (2002) who showed that C decomposition in old, highly weathered soils of tropical forests was strongly constrained by P availability. Overall, these findings indicate that P limitation of microbial process, which until recently has been overlooked, may have profound implications for C and nutrient cycling in forest ecosystems.

4.6 Conclusions

There is a growing body of evidence showing that the presence of particular plant species within a plant community can cause a vast suite of changes in the composition and function of the soil biota, and that these effects have the potential to feedback to the plant in terms of either enhanced or reduced growth rate under those conditions; that is, positive or negative feedback. While empirical evidence of the mechanisms involved is still relatively scarce, there is a growing body of information indicating that such feedback mechanisms, which operate at different spatial and temporal scales, have the potential to act as important drivers of plant community and ecosystem dynamics. At the community level, certain plant species appear to be able to encourage the accumulation of microbes in soil that specifically benefit that plant—for example, due to a change in the availability of nutrients or promotion of mycorrhizal associations—increasing its relative performance and reinforcing its dominance and the loss of other plant types. In contrast, in the case of negative feedback, certain plant species appear to perform worse in their own soil, often due to an accumulation of host-specific pathogens, thereby preventing any single species becoming dominant, enabling more species to coexist. Over long timescales, plant–soil feedbacks also appear to be important drivers of ecosystem productivity; long-term changes in nutrient limitation set feedbacks in motion between decomposers and plant litter, which ultimately determine the productivity and community composition of an ecosystem. While the significance of plant–soil feedbacks is becoming increasingly apparent to ecologists, much is still to be learned about the mechanisms that drive them and also their importance relative to abiotic forces that structure communities and ecosystems.

5 Above-ground trophic interactions and soil biological communities

5.1 Introduction

As shown in the previous chapter, it is increasingly being recognised that to understand how terrestrial ecosystems function requires a combined above-ground–below-ground approach, because of the importance of feedbacks that occur between plants and the decomposer communities. These feedbacks between plants and decomposer organisms are also strongly influenced by other trophic groups, notably above-ground herbivores that are the main consumers of plant material in terrestrial ecosystems, consuming up to 50% of total net primary productivity, depending on ecosystem (Table 5.1) (McNaughton et al. 1989). It has long been recognized that above-ground herbivores are an important driver of ecosystem productivity and vegetation composition (Crawley 1983, 1997; McNaughton et al. 1989), but only recently have ecologists started to consider how they influence decomposer communities, and the consequences of this for ecosystem functioning.

This chapter examines some of the many ways that above-ground herbivores may influence below-ground organisms and processes, and how these changes in soil biological properties may in turn feedback to affect above-ground

Table 5.1 Variability in herbivore consumption (% NPP consumed) between ecosystems. (Adapted from Mc Naughton et al. 1989)

Ecosystem	Consumption (% net primary productivity consumed)
Managed grasslands	30–45
African grasslands	28–60
Herbaceous old fields	1–15
Mature deciduous forests	1.5–2.5

primary production and the grazers themselves. The chapter will highlight the different mechanisms by which herbivores affect decomposers and their activities over various temporal and spatial scales, ranging from the individual plant to the plant community; how these mechanisms operate in different ecosystems will also be discussed.

5.2   Mechanisms

There are a wide variety of ways by which herbivores can influence decomposer organisms and processes, and these mechanisms operate at a variety of temporal and spatial scales, ranging from short-term effects on root exudation at the individual plant level, to long-term effects on soil nutrient cycling at the ecosystem scale, resulting from changes in vegetation composition and resource quality. Three main groups of mechanisms have been identified as being of primary importance (Bardgett and Wardle 2003; Wardle and Bardgett 2004*a*); these are detailed below.

5.2.1   Effects of herbivores on resource quantity

There are two ways by which herbivores can influence the quantity of resources entering the soil. In the short-term, the quantity of resources supplied to soil can be altered through effects of herbivory on plant C allocation and root exudation patterns, whereas in the long-term, the amount of organic matter returned to soil is affected by herbivore-induced shifts in net primary productivity (NPP). Both these mechanisms are intimately linked and collectively influence plant productivity and composition.

Plant allocation and exudation

Plants allocate large proportions of their assimilated C to root exudation (Bokhari 1977), which may stimulate the growth and activity of heterotrophic microbes in the rhizosphere (Whipps and Lynch 1986). Both foliar (Holland et al. 1996; Paterson and Sim 1999; Murray et al. 2004) and root (Yeates et al. 1998; Denton et al. 1999; Grayston et al. 2001*b*) herbivory have been shown to lead to short-term increases in root exudation, which stimulates microbial biomass and C use efficiency by microbes in the rhizosphere (Mawdsley and Bardgett 1997; Denton et al. 1999; Guitian and Bardgett 2000; Hamilton and Frank 2001). In turn, this has been shown to increase the abundance of microbial-feeding fauna (Mikola et al. 2001; Hokka et al. 2004), but reduce AM infection (Gehring and Whitham 1994). Evidence is accumulating to show that these positive effects of herbivory on free-living rhizosphere microbes and higher level consumers in the soil food web can feedback positively to the plant through enhanced soil N availability. Hamilton and Frank (2001), for example, showed that simulated defoliation of the grazing tolerant grass *Poa pratensis* led to increased leaf photosynthesis and root exudation of recently assimilated ^{13}C, which in turn stimulated microbial biomass in the root zone. This in turn increased soil N availability and plant N acquisition, which ultimately benefited plant growth. It was proposed that such mechanisms could explain, in part, the compensatory response of grasses to grazing in high fertility grasslands (Hamilton and Frank 2001) (Fig. 5.1). Similarly, simulated herbivory of tree seedlings has been shown to lead to enhanced N mineralization and inorganic N availability in rhizosphere soil, presumably owing to a stimulation of biological activity resulting from increased rhizodeposition (Ayres et al. 2004). While there is much interspecies variability in the response of plants to defoliation (Guitian and Bardgett 2000; Hokka et al. 2004)—and responses are likely to vary with growth stage of the plant—these findings collectively suggest that physiological responses of plants to herbivory have the potential to stimulate rhizosphere processes that ultimately feedback positively on plant nutrition and plant productivity (Bardgett and Wardle 2003).

Changes in plant productivity

In the long-run, herbivory is often observed to significantly alter NPP, both above- and below-ground. Optimization of above-ground NPP by grazing mammals is well known to occur in grassland ecosystems at intermediate levels of herbivory (McNaughton 1985), and there is theoretical support for promotion of NPP by foliar herbivores occurring at the level of the whole plant community especially in areas of high soil fertility, such as productive grassland ecosystems (De Mazancourt et al. 1999). Similarly, there is evidence that moderate densities of invertebrate herbivory, for example,

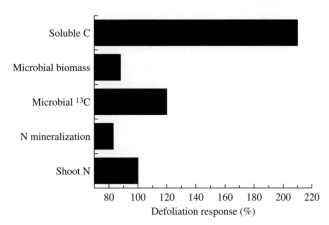

Fig. 5.1   Positive effects of defoliation of the grass *Poa pratensis* on the transfer of recently photosynthesized ^{13}C to rhizosphere microbes, leading to stimulation of microbial biomass, net N mineralization, and plant N acquisition. Data are expressed as % increase relative to undefoliated control. These data indicate that grazed plants stimulate soil microbes and their own N supply. (Data from Hamilton and Frank 2001.)

by grasshoppers, can promote NPP through enhancing soil N cycling (Belovsky and Slade 2000).

There is mixed evidence about how herbivory affects root productivity. Pot experiments consistently show that repeated defoliation reduces root biomass (e.g. Guitian and Bardgett 2000; Mikola et al.2001), and field studies have shown that moose and hare browsing reduces fine root productivity in Alaskan Taiga forest (Ruess et al. 1998). However, fenced exclusion studies on Serengeti grasslands show that mammalian grazers do not necessarily inhibit root biomass and productivity (McNaughton et al. 1998), and in a global literature synthesis Milchunas and Lauenroth (1993) reported both enhancements and reductions in root biomass as a result of herbivore exclusion. There are numerous reports of root herbivores negatively affecting NPP in a range of ecosystems (e.g. Brown and Gange 1990; Ingham and Detling 1990), and also incidences where infestation of roots by insects has led to proliferation in the growth of lateral roots (Brown and Gange 1990).

Herbivore-induced changes in NPP have important knock-on effects to the soil food web and nutrient cycling, but these tend to be quite variable. For example, increasing NPP has been shown to have both positive and negative effects on both microbial biomass and higher trophic levels of the soil food web (Bardgett and Wardle 2003). There are two reasons for such idiosyncratic responses of decomposers to NPP. First, the relative importance of top–down and bottom–up forces in regulating soil food-web components may be context dependent. Second, plants not only provide C resources for microbes but also compete with them for nutrients (Kaye and Hart 1997). Therefore, the direction of effects of herbivore-induced changes in NPP on decomposer organisms and nutrient cycling may be governed by which of the two opposing effects (stimulation of microbes by C addition, inhibition of microbes by resource depletion) dominates. However, despite this uncertainty, it is most likely that positive effects of herbivory on soil biota and nutrient mineralization dominate in sites of high soil fertility, especially in grasslands where C addition to soil from plants typically stimulates soil biological activity since it alleviates C limitation of microbes.

5.2.2 Effects of herbivores on resource quality

Herbivores can induce shifts in the quality of resources entering the soil, by returning dung and urine to the soil and inducing physiological changes in plants. These two mechanisms will be discussed in sequence.

Animal wastes

Ungulate waste and its facilitative effect on soil biological processes is often reported as one of the main driving forces for grazers stimulating N availability and plant nutrient uptake (McNaughton et al. 1997a,b; Frank and Groffman 1998). Herbivorous mammals can return large quantities of undigested and non-assimilated nutrients to the soil as dung, and as assimilated nutrients in urine, and this effectively short-cuts the litter decomposition pathway, providing highly decomposable resources that are rich in labile nutrients, and which can stimulate microbial biomass and activity (Bardgett et al. 1998), net C and N mineralization (Molvar et al. 1993; McNaughton et al. 1997a), and ultimately plant nutrient acquisition and growth. Although the positive effect of animal wastes on soil biological properties is unquestionable, it has been argued that it does not entirely explain the overall positive effects of herbivores on plant productivity at the community scale (Hamilton and Frank 2001). This is because in both agricultural and natural ecosystems, waste patches tend to influence only a relatively small proportion of the surface area (Augustine and Frank 2001).

Much less is known about the effects of waste from invertebrate herbivores on soil processes, although frass (insect feces) is thought to be analogous to ungulate waste in that it is rich in N, with high potential to mineralize on entry into soil. In microcosms, for example, frass from larvae of the gypsy moth (*Lymantria dispar*), which had fed on foliage of *Quercus velutina*, was shown to increase microbial growth and N mineralization in soil (Lovett and Ruesink 1995). While chewing insects return organic matter as frass, sucking insects such as aphids and scale insects often secrete surplus ingested carbohydrates as honeydew, which is an extremely labile C source for microbes when it reaches the ground. Dighton (1978), for example, showed that addition of synthetic honeydew to soils at a rate comparable to that produced by lime aphids (*Eucalipteris tilidae*) on trees of *Tilia* spp. caused an increase in fungal and bacterial populations of 1.3 times and 4.0 times, respectively. Furthermore, honeydew-producing aphids in central European *Picea abies* forests have been shown to cause substantial increases in dissolved organic C in soils, leading to net immobilization of N by soil microflora (Stadler and Michalzik 1998). Given the abundance of honeydew-producing sucking insects in many ecosystems, honeydew effects on decomposer organisms and processes are likely to be of widespread importance (Wardle and Bardgett 2004b).

Litter quality

Foliar herbivory in grasslands of high soil fertility often enhances leaf nutrient concentration (Holland and Detling 1990; Hamilton and Frank 2001) either directly through reallocation of nutrients within individual plants, or indirectly by stimulating soil mineralization processes, as discussed above. Likewise, in forest ecosystems, there are incidences where browsing by mammals has been reported to enhance nutrient concentrations and reduce levels of C-based secondary metabolites, such as phenolics, in foliage (Bryant and Reichart 1992). This has been shown to enhance leaf litter quality, thereby increasing decomposition and mineralization processes in forest ecosystems (Keilland et al. 1997). Stimulation of soil biota and soil processes by foliar herbivory may also be due, in part, to roots of grazed plants producing litter of a higher quality. Seastedt et al. (1988), for example, showed that trimming of shoots of the grass *Andropogon gerardii* increased the N concentration of roots, and proposed that this would stimulate soil organisms through improved resource quality. Likewise, root herbivory by nematodes has been shown to alter the root N content of plants attacked by nematodes (Bardgett et al. 1999a). Such herbivore-induced increases in foliage and root nutrient concentration will invariably have a positive influence on soil biota and soil mineralization processes.

Herbivores can also induce the production of secondary plant compounds in foliage and roots, which negatively impact on soil biota owing to reduced litter quality. For example, severe defoliation of trees, such as that caused by periodic invertebrate attack, often results in reduced concentrations of N and increased concentrations of certain secondary metabolites (e.g. phenolics) in subsequently produced foliage. It is unclear whether this response is 'induced' for plant defence against browsing (Rhoades 1985) or whether phenolic production is a physiological response of trees recovering from the nutritional deficiency associated with defoliation (Bryant et al. 1993). Whatever the mechanism responsible, the net result is likely to be the production of leaf litter with characteristics that are less favourable for decomposer organisms. There is no substantial direct evidence of this, but Findlay et al. (1996) showed that cellular damage caused by spider mites to seedlings of *Populus deltoides* increased the concentration of polyphenols in foliage, resulting in a 50% reduction in the decomposition rate of subsequently produced leaf litter. The importance of these events in nature is not known, but presumably they would negatively influence soil nutrient mineralization and hence plant productivity.

5.2.3   Effects of herbivores on vegetation composition

Over long timescales, herbivory often leads to significant changes in the functional composition of the plant community, leading to changes in both the quantity and quality of litter inputs to soil, which in turn affects soil biota and soil nutrient cycling. Many grazers are selective in their feeding habit, preferentially grazing on palatable species, often of high nutrient content. Depending upon context, such selective feeding by grazing mammals can either accelerate or retard vegetation succession (Fig. 5.2), with important implications for soil biological properties (Bardgett and Wardle 2003). Retardation of succession occurs when later successional plant species are suppressed as a result of their sensitivity to grazing (Van der Wal et al. 2000) and/or when dominant plant species benefit from herbivory, for example, through compensatory growth and positive feedbacks between plants and herbivores (Augustine and McNaughton 1998). As discussed above, this is

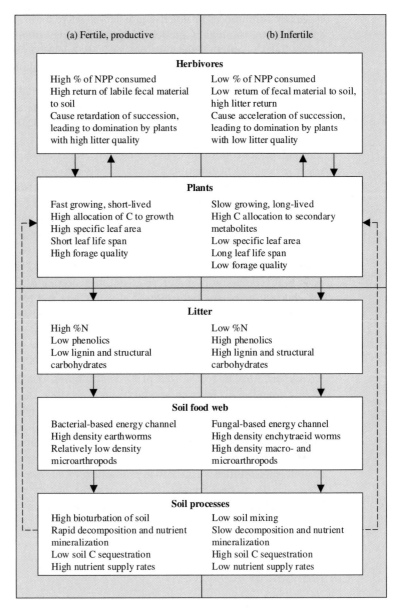

Fig. 5.2 The difference in fundamental plant traits between species that dominate (a) fertile systems which support high herbivory, and (b) infertile habitats which support low herbivory, functions as a major ecological driver. This occurs through plant traits affecting the quality and quantity of resources that enter the soil, and the key ecological processes in the decomposer subsystem driven by the soil biota. These linkages between below-ground and above-ground systems feedback (dotted line) to the plant community positively in fertile conditions (a) and negatively in infertile ecosystems (b). (Redrawn with permission from Wardle et al. 2004a, Science 304, 1629–1633. Copyright 2004 AAAS)

especially common in productive grasslands where herbivory has positive effects on soil biological properties by preventing colonization by later successional plants that produce poorer litter quality, as well as through returning C and nutrients to the soil in labile forms as dung and urine.

Herbivores accelerate succession when earlier successional plant species—which tend to be more palatable and nutrient-rich than later successional species (Cates and Orians 1975; Grime 2001; Wardle 2002)—are disadvantaged by selective grazing. This effect of herbivory appears to be especially common under conditions of low soil fertility, for example, in forests where selective feeding by mammals leads to replacement of palatable plant species by those which are better defended against herbivory, but produce litter of poorer quality. A classic example of this is that of Pastor et al. (1988, 1993) who found in Boreal forest that browsing by moose on deciduous tree species with nutrient-rich foliage increased the dominance of a coniferous tree *Picea* spp., which is of low palatability and produces litter of poorer quality. This shift in vegetation composition resulted in reduced soil microbial biomass and slower rates of litter decomposition and soil N mineralization, thereby reducing soil N availability and plant productivity. Similarly, insect and mammalian herbivory in N-limited oak savannah was shown by Ritchie et al. (1998) to greatly decrease the abundance of plant species with N-rich tissue (especially the legume *Lathyrus venosus*), thereby reducing the positive contribution of these plants to N cycling in these ecosystems. Also, in synthesized plant communities, Buckland and Grime (2000) found that invertebrate herbivores, notably the slug *Deroceras reticulatum*, suppressed early successional fast-growing plant species when grown under infertile conditions; however, under fertile conditions early successional species dominated when herbivores were present, presumably because of their ability to compensate for tissue loss through defoliation. Overall, these studies suggest that promotion of succession by herbivory adversely affects decomposers by reducing the quality of litter entering the soil, while suppression of succession by herbivores should have the opposite effect.

5.3   Comparison of ecosystems

From the above, it is apparent that there are many ways by which herbivores can strongly influence soil biological properties, and these mechanisms can have either positive or negative consequences for ecosystem productivity. The variety of mechanisms is consistent with the idiosyncratic and heterogeneous nature of herbivore impacts that are commonly observed in the field and reported in literature (Bardgett and Wardle 2003). Understanding the effects of herbivores on soil biological properties and ecosystem function is therefore strongly context dependent, with different mechanisms dominating in different ecosystems. This is now illustrated by comparing effects of large mammalian herbivores on soil biological properties in a range of ecosystems, namely grasslands, tundra, and forests.

5.3.1 Effects of herbivores on soil and ecosystem properties of grasslands

Grasslands, including steppes, savannas, and prairies are important terrestrial ecosystems covering about a quarter of the Earth's land surface. Grasslands (Fig. 5.3) tend to support much higher mammalian herbivore biomass than do other ecosystems, and across grasslands the biomass of herbivores increases with increasing NPP (McNaughton 1985; Oesterheld et al. 1992), and is especially high on managed grasslands due to controls on predation, disease, and parasitism (Osterheld et al. 1992). Positive effects of herbivory on soil biological properties and plant productivity tend to dominate in these situations, especially in more productive sites where herbivores often acquire most of their food (Fig. 5.4). Here, a key mechanism for positive effects of herbivory is that nutrients are returned to the soil in labile forms (e.g. dung and urine, and high quality litter) leading to

Fig. 5.3   Effects of grazing by large herbivores (sheep) on grassland ecosystems as studied using fenced exclosures erected in 1956 in Llyn Llydaw, Snowdonia, Wales. On the inside of the fence the vegetation is dominated by dwarf-shrubs such as *Calluna vulgaris* and *Vaccinium myrtillus*, with occasional regenerating trees, which are eliminated by grazers and replaced by grasses outside the exclosure. (Image by Richard Bardgett.)

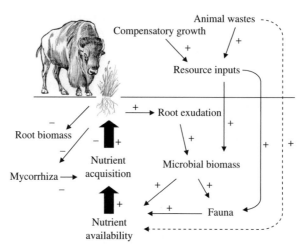

Fig. 5.4   Schematic diagram of the effects of herbivory on producer–decomposer feedbacks in fertile grasslands that result from changes in the quantity of resources returned to the soil. These mechanisms are most common in nutrient-rich grasslands, where dominant plant species benefit from herbivory through positive feedbacks between herbivores, plants, and soil biota, and through prevention of colonization by later successional plants that produce poorer litter quality. (Redrawn with permission from Ecological Society of America; Bardgett and Wardle 2003.)

positive effects on soil biota and nutrient mineralization processes, and hence plant productivity (Bardgett and Wardle 2003). Additionally, plant physiological responses to herbivory, such as enhanced root exudation, also stimulate soil biota and nutrient mineralization processes in grassland ecosystems, leading to enhanced nutrient supply to plants (Guitian and Bardgett 2000; Hamilton and Frank 2001). These mechanisms, together, reinforce soil fertility and ultimately benefit plant productivity. As a result, soils of grazed grasslands tend to have small amounts of dead litter on the soil surface but have large amounts of organic N and C (Bardgett and Cook 1998). These features combine to produce a soil environment that sustains an abundant and diverse faunal and microbial community (Bardgett and Cook 1998).

There are many examples of such positive effects of herbivores on soil biological properties and plant productivity in grassland ecosystems. For example, in the Serengeti grasslands, large, free-ranging mammalian grazers have been shown to accelerate soil N cycling, alleviating nutrient deficiency, and enhancing their own carrying capacity (McNaughton et al. 1997a). This positive effect of herbivory on N cycling has been attributed to the mechanisms identified above, and also to enhanced activity of soil ureases, microbial enzymes responsible for the hydrolysis of urea, which increased with the level of grazer activity (McNaughton et al. 1997b). Similarly, mammalian grazers have been shown to enhance rates of N mineralization in native grasslands in the Yellowstone National Park (Tracey and Frank 1998), despite not affecting the size of the soil microbial biomass that was more strongly influenced by topography. In British upland grasslands, cattle and sheep are known to increase soil microbial activity and N cycling and hence plant production (Floate 1970a,b; Bardgett et al. 1997).

Mammalian grazers have also been shown to have significant effects on the abundance and composition of soil biotic communities. For example, studies of temperate grasslands in the UK have shown that grazing by sheep increases microbial biomass and activity, and also the abundance of microbial consumers, namely nematodes and microarthropods (Bardgett et al. 1993a, 1997). The structure of microbial communities is also affected by the long-term history of grazing by sheep in these temperate grasslands. As shown by Bardgett et al. (2001b) microbial biomass of soil is maximal at low-to-intermediate levels of grazing and the phenotypic evenness (a component of diversity) of the microbial community declines as intensity of grazing increases. These authors also showed that bacterial-based energy channels of decomposition dominate communities of heavily grazed sites, whereas in systems that are less intensively grazed, or completely unmanaged, fungi have a proportionally greater role (Bardgett et al. 2001b). This finding is consistent with studies of effects of grazing in US grasslands, which show that the abundance of bacterial-feeding nematodes is greater in grazed than ungrazed annual and prairie grassland (Freckman et al. 1979; Ingham and Detling 1984), and other studies of UK grasslands that show a shift from fungal towards bacterial dominance within the microbial community as grazing intensity increases (Bardgett et al. 1993a; Grayston et al. 2001a, 2004).

While positive effects of grazing on soil biota and processes appear to dominate in productive grassland systems, negative effects are also reported. For example, increased sheep stocking density on an Australian pasture (10, 20, 30 sheep ha^{-1}) severely reduced numbers of Collembola in the surface soil (King and Hutchinson 1976; King et al. 1976). Likewise, reductions in Collembola numbers were associated with increased sheep stocking density of a lowland perennial ryegrass (*Lolium perenne*) grassland (Walsingham 1976). These responses were attributed to changes in soil pore space and surface litter, both of which were greatly reduced with increased sheep grazing intensity. More recently, Sankaran and Augustine (2004) showed that removal of mammalian grazers for two years by fencing in a semiarid grassland in Kenya actually increased soil microbial biomass in sites of both high and low fertility, despite the fact that grazing stimulated above-ground plant production in nutrient-rich sites and depressed it in nutrient-poor sites (Fig. 5.5). This suggests that grazers have consistently negative effects on soil microbial biomass in these grasslands, which has been attributed to: (1) reductions in plant C input to soil resulting from a depression in root production; and (2) the diversion of plant C to herbivore respiration and

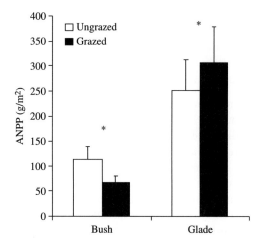

Fig. 5.5   Effects of grazing by large herbivores on above-ground production (g/m²over 5 months) in glade (nutrient-rich) and bush (nutrient-poor) communities in Kenya. Values are means with standard errors. Error bars show + 1 SE based on among-block variation. * indicates significant differences at the α = 0.05 level. (Redrawn with permission from Ecological Society of America; Sankaran and Augustine 2004.)

growth in these heavily grazed ecosystems. Another important finding of this study was that microbial biomass was highly correlated with soil C content across all sites studied, suggesting that landscape-scale constraints on soil organic matter content overarch grazer effects on microbial abundance (Fig. 5.6). This finding is also consistent with studies of other grass dominated ecosystems, that show that while grazers have significant effects on soil biological properties, they are often out-weighed by higher order landscape controls on these ecosystem properties (Tracey and Frank 1998; Harrison and Bardgett 2004).

5.3.2   Effects of herbivores on soil and ecosystem properties of arctic tundra

It is generally thought that the ability of herbivores to increase decomposition and nutrient mineralization processes, thereby sustaining the production of preferred forage plants, may be of particular importance in strongly resource-limited environments such as arctic tundra where herbivore food supplies are low (Jefferies 1988). There is a growing body of evidence indicating that tundra ecosystems are indeed strongly influenced by grazing animals, notably by reindeer (Manseau et al. 1996; Jano et al. 1998; Virtanen 2000). Reindeer feed on lichens (notably *Cladina* spp.) throughout the Arctic and heavy grazing by reindeer is known to cause dramatic changes in tundra vegetation composition, causing a shift from moss and lichen rich heath into grass dominated meadow (Olofsson and Oksanen 2002; Olofsson et al. 2004; Van der Wal and Brooker 2004). These changes in vegetation composition

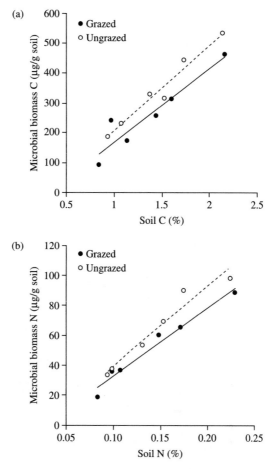

Fig. 5.6 Relationships between microbial biomass C and soil organic C (a) and microbial biomass N and soil total N (b) in plots in grazed and ungrazed areas in glade and bush communities, Kenya. Closed circles represent grazed areas and open circles ungrazed sites. Equations for regression lines: $MBC_{grazed} = 245.7$ (Soil C) $- 77.2$ ($r^2 = 0.89$, $P = 0.004$); $MBC_{ungrazed} = 290.4$(Soil C) $- 85.1$ ($r^2 = 0.96$, $P < 0.001$); $MBN_{grazed} = 455.4$ (soil N) $- 12.7$ ($r^2 = 0.97$, $P < 0.001$); $MBN_{ungrazed} = 529.1$ (Soil N) $- 13.5$ ($r^2 = 0.95$, $P = 0.001$). (Redrawn with permission from Ecological Society of America; Sankaran and Augustine 2004.)

have cascading effects on soil biological properties, but as in other ecosystems their nature is highly site-specific. For example, while reductions in soil microbial biomass by reindeer grazing are frequently observed, especially in nutrient-poor heathlands (Väre et al. 1996; Stark et al. 2000, 2003; Stark and Grellman 2002), effects of reindeer on microbial biomass N and rates of N mineralization are highly variable, with both increases and decreases in N mineralization being reported, depending on context (Stark et al. 2000; 2003; Stark and Grellman 2002; Olofsson et al. 2004; Van der Wal et al. 2004).

Reindeer grazing also appears to have mixed effects on plant litter decomposition rates; while Stark et al. (2000) found decomposition rates of litter to be impaired when placed outside exclosure plots, Olofsson and Oksanen (2002) found the reverse trend.

As in other ecosystems, the effects of reindeer on soil biological properties vary as different mechanisms by which they affect decomposers operate differently in different contexts. Negative effects on soil biological properties most probably arise owing to reindeer removing the protective cover of lichens and exposing the soil biota to a less favourable microclimate (Fig. 5.7) (Stark et al. 2000), and through reindeer trampling plant roots

Fig. 5.7   Heavy grazing by reindeer in Boreal forest in Sweden has removed the protective cover of lichens from the forest floor, exposing the soil biota to a less favourable microclimate. (Image by Richard Bardgett.)

and reducing C inputs to the soil (Stark et al. 2003). Positive effects on soil biological properties and nutrient availability may arise through reindeer promoting plant species that produce better quality litter, by increasing soil temperature, and returning labile dung and urine materials to the soil (Olofsson et al. 2004). Indeed, studies in the high Arctic, Spitsbergen (Fig. 5.8) have shown that reindeer grazing promotes soil N availability and plant productivity mainly due to effects of faeces (Van der Wal et al. 2004). In this study, experimental faecal addition to moss dominated tundra was shown to increase grass growth and microbial biomass in soil, but lead to a reduction in the depth of the moss layer influencing soil temperature (Fig. 5.9). This took some three years to occur and the cause for this was thought to be a direct fertilizing effect of faeces in these strongly nutrient-limited situations, and also a suppressive effect of faeces on the depth of the moss layer, which strongly regulates soil temperature owing to its ability to hold moisture (Brooker and Van der Wal 2003). Since the greatest reduction in moss depth occurred where fouling increased microbial biomass the most, it was proposed that enhanced decomposition of moss by a more abundant microbial community could be the cause (Van der Wal et al. 2004). Overall, these findings support the notion that reindeer grazing is of fundamental importance to tundra ecosystem productivity, supporting the hypothesis that plant–soil feedbacks resulting from herbivory are instrumental in promoting grass growth while suppressing mosses (Fig. 5.10).

Fig. 5.8 Reindeer grazing on moss dominated ecosystems in Svalbard, high Arctic. (Image by Richard Bardgett.)

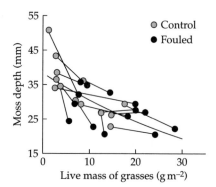

Fig. 5.9   Relationships between depth of the moss layer and live mass of grasses four years after plots received additional faeces (fouled plots) and their associated untreated controls. Individual pairs of control and fouled plots are connected. The thick black line indicates the overall relationship between moss depth and grass biomass. (Redrawn with permission from the Nordic Society Oikos; Van der Wal et al. 2004.)

The finding that reindeer grazing promotes grass growth at the expense of mosses is consistent with the keystone herbivore hypothesis of Zimov et al. (1995). These authors proposed that mega-herbivore extinction in Alaska and Russia 10,000–12,000 years ago, rather than changes in climate alone, have led to vegetation transforming from grazed steppe grassland to wet moss tundra (Zimov et al. 1995). This would have in turn led to domination by plant species that produce poorer litter quality, greater waterlogging and reduced soil temperature, and retardation of soil N mineralization, providing a feedback that maintained the dominance of the tundra vegetation.

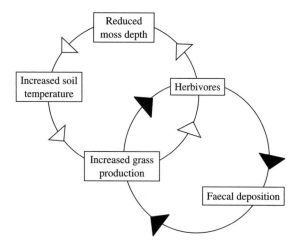

Fig. 5.10   Schematic diagram showing cascading effects of reindeer grazing on moss dominated high arctic ecosystems. (Adapted from Van der Wal et al. 2004.)

5.3.3 Effects of herbivores on soil and ecosystem properties of forests

As already noted, herbivores are highly selective in their feeding and in forests they show a clear preference for more palatable deciduous species over conifer species. As a result, browsing in forests often leads to a decline in the abundance of palatable species, and an increase in the dominance of less palatable species that also produce poor quality litter. A classic example of this is the already mentioned study of the effects of moose browsing on soil nutrient cycling in Boreal forest on Isle Royale National Park, Michigan, by Pastor et al. (1988, 1993) (Fig. 5.11). Using 40-year-old fenced exclosures, these authors showed that selective browsing of deciduous forest species by moose increased the frequency of spruce, which produces litter of very low quality, at the expense of more palatable deciduous tree species. This in turn led to reduced soil microbial biomass and activity, and reduced N mineralization . While moose pellets had beneficial effects on soil nutrient availability, this was not sufficient to offset the depression in N and C mineralization in soil resulting from the increased abundance of unbrowsed

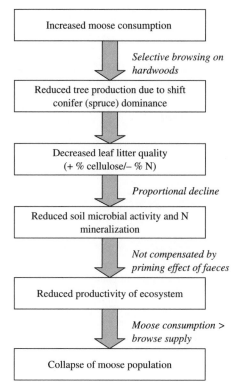

Fig. 5.11 Schematic diagram of the degenerative feedback loop caused by moose browsing on Isle Royale National Park, Michigan. (Pastor et al. 1988, 1993.)

spruce. It was concluded that, in the long-run, high rates of moose browsing depress N mineralization and NPP through the indirect effects on recruitment into the tree stratum, and subsequent depression of litter N return and litter quality. This suggests that the effects of herbivores in forest ecosystems may be amplified by positive feedbacks between plant litter and soil nutrient availability.

Similarly, other studies reveal strong negative effects of browsing by large mammalian herbivores in forests. A study of the effects of red deer browsing (Fig. 5.12) on nutrient cycling in regenerating forests in the Scottish

Fig. 5.12   The removal of red deer browsing at Creag Meagaidh National Nature Reserve in the Scottish Highlands, has led to a four-fold increase in net N mineralization, which feeds back positively on the regeneration of birch trees in this ecosystem. (Image by Richard Bardgett.)

Highlands revealed that while topographic factors were the dominant factors affecting soil biological properties in this ecosystem, browsing significantly reduced C and N mineralization, potentially reducing ecosystem productivity (Harrison and Bardgett 2004). This effect was attributed to a negative effect of browsing on growth of the dominant tree species, *Betula pubescens*, which is well known to have positive effects on soil nutrient cycling through improvements in the quality of litter inputs to soil and through the action of its roots (Harrison and Bardgett 2003). As in the study of Pastor et al. (1988, 1993) negative effects of browsing on nutrient cycling related to tree growth were not compensated by the priming effects of red deer faeces, which in this ecosystem will have been highly localized.

While many forest ecosystems support native herbivore populations, they are increasingly being colonized by invasive herbivorous mammals in many parts of the world. In New Zealand, for example, several species of large browsing mammals, including the European red deer and feral goats, were introduced by humans between the 1770s and 1920s. Prior to this, browsing mammals did not exist in New Zealand. This has caused widespread shifts in the forest understorey throughout the country, often reducing or eliminating broad-leaved fast-growing palatable plant species and promoting unpalatable fern and monocotyledonous species. The impacts of this on soil biological properties were studied by Wardle et al. (2001) using a number of long-term fenced exclosure plots that were erected in indigenous forests in the 1950s. Measurements made inside and outside each exclosure revealed that browsing mammals had consistent adverse effects on the density and diversity of vegetation present in the understorey, and tended to promote unpalatable species with poor litter quality (typically monocotyledonous and fern species, and small-leaved dicotyledonous species) at the expense of palatable species with higher litter quality (typically large-leaved dicotyledonous species). Despite the consistency of these trends, the response of the soil biota was far less predictable (Wardle et al. 2001), presumably reflecting the variety of mechanisms by which browsers affect decomposers at different locations. Most groups of smaller-bodied soil organisms (notably microfauna and microflora) showed idiosyncratic responses to browsing, with both strong positive and strong negative effects of browsing mammals occurring depending upon the site considered. Ecosystem processes and properties driven by the soil biota, such as soil C mineralization, and C and N sequestration in the soil, also showed these types of context-dependent responses. In contrast, animals with larger body sizes such as soil microarthropods and macrofauna were consistently adversely affected by browsing mammals, probably because these organisms are more susceptible than smaller ones to physical disturbances (e.g. trampling) caused by mammalian herbivores.

As in other ecosystem types, the effect of browsing animals on soil biological properties in forests is highly context dependant, although there is strong evidence to suggest that changes in vegetation structure caused by browsing in unproductive forest systems, such as Boreal forests, can lead to negative effects on soil biota and nutrient cycling, thereby negatively affecting forest productivity. However, it is evident, as in other ecosystems, that the effect of large mammalian herbivores on these properties will vary depending on a host of site specific factors, including the intensity and history of herbivory, site fertility, and vegetation structure.

5.4   Conclusions

Understanding the roles of soil biota in ecosystems requires that they be studied in the context of interactions that occur between the above-ground and below-ground world. From this chapter, it is evident that these interactions are significantly affected by organisms that feed on plants, and that these responses in part determine the response of the ecosystem to herbivore pressure. This chapter has identified three key mechanisms by which herbivores can indirectly affect decomposer organisms and soil processes by altering the quantity and quality of resources that enter the soil. These mechanisms operate either in the short-term, by involving physiological responses of individual plants to herbivore attack, or in the long-term, involving alteration of plant productivity and vegetation com-position, and subsequent changes in the quantity and quality of litter inputs to soil. Owing to the variety of possible mechanisms, the effects of foliar and root herbivores on soil biota and ecosystem function are idiosyncratic and hence difficult to predict, with positive, negative, or neutral effects of herbivory being possible depending upon the balance of these different mechanisms. The variety of mechanisms is consistent with the idiosyncratic and heterogeneous nature of herbivore impacts that are commonly observed in the field.

One important conclusion that does emerge from literature is that the magnitude of herbivore effects on soil biota and soil processes, and the con-sequence of this for producer–decomposer feedbacks, differs greatly across ecosystems, and this appears to depend largely on soil fertility and on the proportion of NPP that is consumed by herbivores. This proportion varies considerably across ecosystems, and in more productive ecosystems, which are dominated by palatable, nutrient-rich plants, and support a greater diversity and level of herbivory, positive effects of herbivory on soil biota and soil processes appear to dominate. This is largely on account of the priming effects of animal wastes and also plant physiological responses to herbivory, such as enhanced root exudation, which ultimately enhance nutrient supply to plants. These mechanisms, together, reinforce soil fertility and ultimately benefit plant productivity at the ecosystem scale. In contrast, in unproductive ecosystems of low soil fertility, with low herbivore consumption rates, negative effects on soil biota and soil processes appear to dominate. Here, selective grazing leads to changes in the functional composition of vegetation, and especially the dominance of defended plant species that produce litter that is of poor nutritional quality to decomposers. The net effect of this is low levels of soil biotic activity, nutrient mineralization, and supply rates of nutrients from soil, and hence a reduction in plant productivity at the ecosystem scale.

6 Soil biological properties and global change

6.1 Introduction

Human activities have a substantial and ever growing influence on ecosystems to the extent that much of the Earth's land surface has been totally transformed by a suite of global change phenomena (Vitousek et al. 1997). The most obvious influence of humans on the Earth is through the increasing transformation of land for agriculture and forestry, which at present represents the primary driving force for the loss of biological diversity worldwide (Wilson 2002). In addition, every terrestrial ecosystem on Earth is now affected by a suite of other global change phenomena, namely the alteration of climate through atmospheric CO_2 enrichment, invasions of alien species into new territories, and increasing rates of N deposition (Fig. 6.1). As a result, most aspects of the structure and function of ecosystems, including the biological communities and functioning of soils, cannot be understood without considering the strong, often dominant and multiple influences of humans.

While the significance of global change phenomena for ecosystem perform-ance is widely recognized and discussed, relatively little is known about the actual mechanisms that drive ecosystem responses to them. It has been argued that to understand impacts of global change on ecosystems and the mechanisms that underlie them requires explicit consideration of linkages between above-ground and below-ground biota (Wardle et al. 2004a). This is because the overall effect of global change phenomena on the structure and function of terrestrial ecosystems is likely to relate to their indirect effect on soil biota and the processes that they drive. This chapter explores this issue, discussing how particular global change phenomena impact on soil biota and their activities, and how these effects feedback to nutrient dynamics and the productivity and structure of above-ground communities. Earth's

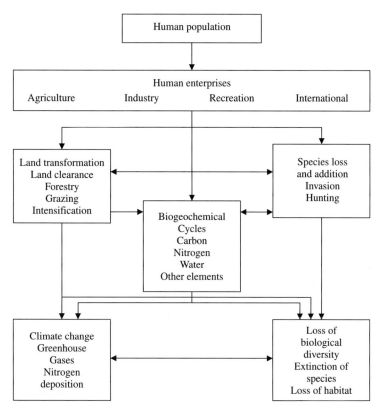

Fig. 6.1   Schematic diagram of humanity's direct and indirect effects on the Earth system. (Redrawn with permission from Vitousek et al. 1997, Science 277, 494–499. Copyright 1997, AAAS)

ecosystems are subject to multiple and simultaneous assaults of global change phenomena; this chapter, however, will take selected global change phenomena, namely climate change, N deposition, invasive species, and land use change, to explore how single components of global change impact on soil and ecosystem processes.

6.2   Climate change

Rising atmospheric concentrations of CO_2 represent the clearest and most documented signal of human alteration of the Earth system (Vitousek et al. 1997). Since pre-industrial years (pre-1800), atmospheric CO_2 concentrations have risen by some 30% from 280 to 365 ppm, due mostly to dramatic increases in the burning of fossil fuels and clearance of primary lands for agriculture. By the end of the century—depending on future industrial trends—concentrations are predicted to reach 540–970 ppm. The general view among climatologists is that this rise in CO_2 has already substantially affected the world's climate, and according to predictions in the Third Assessment Report of the Intergovernmental Panel on Climate Change (IPCC 2001), it will force further change in coming years, leading to a rise in average temperatures and a greater occurrence of extreme weather events. This section will consider how both these factors, namely rising CO_2 and temperature, independently impact on soil biota and the processes that they drive, and ultimately how climate change will influence the ability of soils to store C.

6.2.1 Elevated CO_2 and soil biota

Effects of elevated atmospheric concentrations of CO_2 on soil biota and processes are most likely to be indirect, via changes in the productivity and composition of plant communities, which influence both the amount and quality of C inputs to soil. (Direct effects of elevated CO_2 are generally thought to be of little significance for soil biota, due largely to the high concentration of CO_2 commonly encountered in pore spaces of soil.) Many studies have shown that rising atmospheric CO_2 has the potential to increase plant photosynthesis, leading to increased plant productivity (reviewed by Curtis and Wang 1998). However, positive effects of CO_2 enrichment on plant biomass appear to occur mostly under nutrient-rich conditions, especially when N is not limiting (Curtis and Wang 1998). This suggests that the response of many semi-natural and natural ecosystems—which are typically N limited—to elevated CO_2 will be constrained by nutrient limitation (Oren et al. 2001; Lou et al. 2004).

Under nutrient replete conditions where elevated CO_2 does increase plant growth, it is likely that the return of plant-derived C to soil will also increase, stimulating soil microbial growth and activity. This has been shown to be the case in a number of pot experiments where radioactive tracers (^{14}C) have been used to trace increases in the flux of photoassimilate C through roots to soil under elevated CO_2 (Billes et al. 1993; Paterson et al. 1996; Rouhier and Read 1999). While it has been proposed that such increases in the supply of root-derived C to soil will increase the growth and activity of soil microbes (Zak et al. 1993), evidence for this is very mixed, with a cocktail of positive and neutral responses being reported (Table 6.1). Reports on the effects of elevated atmospheric CO_2 on microbial community structure are also very mixed, with some studies showing no response (e.g. Zak et al. 1996, 2000; Kampicheler et al. 1998; Montealegre et al. 2002; Ronn et al. 2002) and others reporting significant increases in the abundance of certain microbial groups, such as decomposer fungi (Klamer et al. 2002; Philips et al. 2002) and AM fungi under conditions of low nutrient availability (Klironomos et al. 1997; Staddon et al. 2004). This discrepancy in effects of elevated CO_2 could be owing to a variety of factors, but it is most likely a result of differences in the degree to which

Table 6.1 Effects of elevated atmospheric CO_2 concentrations on microbial biomass C and N

Study description[a]	Species/community	Group[b]	Microbial		Reference
			C	N	
342, 692 ppm / <1 yr / OTC	Bigtooth aspen (Populus grandidentata)	W	↑		(Zak et al. 1993)
357, 707 ppm / 3 yrs / OTC	Trembling aspen (Populus tremuloides)	W		0	(Zak et al. 2000)
367, 715 ppm / 2 yrs / OTC	Trembling aspen (P. tremuloides)	W	0	0	(Mikan et al. 2000)
Ambient, 535 ppm / 3 yrs / FACE	Trembling aspen (P. tremuloides)	W	0		(Larson 2000)
	Trembling aspen (P. tremuloides) and birch (Betula papyrifera)	W			
375, 700 ppm / 2 yrs / GH	Yellow birch (Betula alleghaniensis)	W		0	(Berntson and Bazzaz 1997)
410, 757 ppm / 2 yrs / OTC	Scrub oak	W	0	0	(Schortemeyer et al. 2000)
Ambient, +350 ppm / 1 yr / OTC	Scrub oak	W		0	(Hungate et al. 1999)
340, 690 ppm / <1 yr / OTC	Model poplar system	W			(Lussenhop et al. 1998)
Ambient, 200 ppm / 2 yrs / FACE	Loblolly pine (Pinus taeda)	W	0	0	(Allen et al. 2000)
365, 565 ppm / 4 yrs / FACE	Loblolly pine (P. taeda)	W	0	0	(Finzi et al. 2002)
Ambient, 600 ppm / 5 yrs / OTC	Subartic heath	W/G/H	0	0	(Johnson et al. 2000)
Ambient, 2×ambient / 4 yrs / OTC	Tallgrass prairie	G/H	0/↑[c]	0/↑[c]	(Rice et al. 1994)
Ambient, +350 ppm / <1 yr / OTC	Sandstone grassland	G/H	↑		(Hungate et al. 1997a)
	Serpentine grassland	G/H	↑		
Ambient, 720 ppm / 3 yrs / OTC	Sandstone grassland	G/H	↑		(Hungate et al. 1997b)
376, 700 ppm / decades / natural CO_2 vent	Grassland	G/H	↑	↑	(Ross et al. 2000)
350, 700 ppm / <1 yr / GC	Tall herb community	G/H	↑	↑	(Diaz et al. 1993)
	Grassland	G/H	↑	↑	
355, 680 ppm / 4 yrs / OTC	Grassland	G/H	0	0	(Niklaus and Korner 1996)
Ambient, +200 ppm / <1 yr / GC	Model grassland	G/H	0	0	(Kampichler et al. 1998)
350, 525, 700 ppm / 2 yrs / GC	Grassland	G/H	0	0	(Newton et al. 1995)
Ambient, 710 ppm / <1 yr / OTC	Wild oat (Avena fatua)	G		0	(Hungate et al. 1996)
	Soft brome (Bromus hordeaceus)	G		0	
	Italian ryegrass (Lolium multiflorum)	G		0/↓[d]	
	(Vulpia michrostachys)	G		0/↓[d]	
	Goldfields (Lasthenia californica)	H		0	
	Dotseed plantain (Plantago erecta)	H		0	

Table 6.1 (*Continued*)

Study description[a]	Species/community	Group[b]	Microbial		Reference
			C	N	
390, 690 ppm / <1 yr / GC	Perennial ryegrass (*Lolium perenne*)	G	↑		(Schenk et al. 1995)
	White clover (*Trifolium repens*)	H	↑		
350, 600 ppm / 2 yrs / FACE	Perennial ryegrass (*L. perenne*)	G	0		(Schortemeyer et al. 1996)
	White clover (*T. repens*)	H	0		
350, 700 ppm / <1 yr / GC	Perennial ryegrass (*L. perenne*)	G		0[e]	(Gorissen and Cotrufo 1999)
	Common Bent (*Agrostis capillaris*)	G		0[e]	
	Sheep's fescue (*Festuca ovina*)	G		0[e]	
350, 700 ppm / <1 yr / GH	Compact brome (*Bromus madritensis*)	G	↑		(Dhillion et al. 1996)
369, 600 ppm / <1 yr / GC	Barley (*Hordeum disticum*)	G		↑[e]	(Martin-Olmedo et al. 2002)
450, 720 ppm / <1 yr / GC	Bermuda grass (*Cynodon dactylon*)	G	0	0	(Paterson et al. 1996)

[a] Atmospheric CO_2 concentrations/experimental duration/design (FACE, free air CO_2 enrichment; OTC, open top chamber; GH, greenhouse; GC, growth cabinet).

[b] Plant species type (W, woody; G, graminoid; H, herbaceous).

[c] No change with low N, increased with high N in one year; unaffected in another year.

[d] No change with high N, decreased with low N.

[e] Water soluble N.

Notes: ↑, 0, ↓ represent increased, no change and decreased, respectively.

components of the microbial community are regulated by top-down (i.e. predation) versus 'bottom-up' forces (i.e. resource supply); any potential increase in microbial biomass, or the abundance of a particular microbial group, could be dampened by increased consumption by microbial-feeding soil animals (Wardle 2002). With this in mind, considerable care must be taken in predicting how soil microbial communities will behave in complex natural ecosystems under elevated CO_2.

Where changes in microbial biomass and community structure do occur in response to elevated CO_2, they have the potential to influence higher trophic levels. This was shown to be the case by Jones et al. (1998) in a study of model grassland ecosystems—the Ecotron—where elevated CO_2 increased below-ground inputs of C, leading to changes in the relative abundance of decomposer fungi, which in turn changed the abundance and structure of the fungal-feeding collembolan community (Fig. 6.2). On the whole, however, responses of soil fauna to elevated CO_2 are as conflicting and context-dependent as they are for their microbial prey (Wardle 2002). For example, the abundance of nematodes has been shown to be unaffected by elevated CO_2 in soil planted with pine (Markkola et al. 1996) or trembling aspen (Hoeksema et al. 2000), but increased under raised CO_2 in experimental grassland turves after 490 days (Yeates et al. 1997*b*), but not after 261 days (Newton et al. 1995). Population densities of enchytraeid

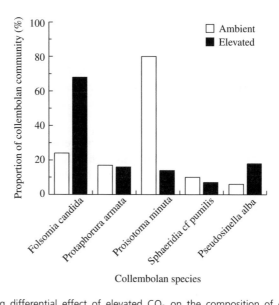

Fig. 6.2   Data showing differential effect of elevated CO_2 on the composition of collembolan communities in soil of model grassland ecosystems, the Ecotron. Data are relative abundances (%) of five species of Collembola after 9 months (3 plant generations) of ambient (open bars) and elevated CO_2 (shaded bars) (means ±SE). (Data from Jones et al. 1998.)

worms were found to increase in response to elevated CO_2 in grassland (Newton et al. 1995; Yeates et al. 1997*b*), but declined in pot experiments planted with pine (Markkola et al. 1996). There is also little consistency in the response of microarthropods to elevated CO_2. For example, Rillig et al. (1999) found that microarthropod abundance increased in two grasslands in response to elevated CO_2, whereas in a pot experiment, Markkola et al. (1996) found no effect of this treatment on collembolans in soil under pine. Such lack of consistency suggests that general predictions on the effects of elevated CO_2 on soil food webs are difficult to make.

6.2.2 Influence of elevated CO_2 on soil nutrient availability

It would be reasonable to expect that changes in microbial biomass and/or community structure that result from CO_2 enrichment of the atmosphere would influence rates of decomposition, nutrient mineralization, and nutrient availability to plants. In view of the contrasting effects of elevated CO_2 on soil food webs, it is perhaps unsurprising that effects on soil nutrient cycling are equally mixed (Table 6.2). For example, Zak et al. (1993) showed that elevated CO_2 enhanced plant photosynthesis and the supply of root-derived C to soil, leading to increased microbial biomass and N mineralization, and uptake of N by plants (Fig. 6.3). These findings were taken to suggest that elevated atmospheric CO_2 concentrations could have a positive feedback effect on soil C and N dynamics promoting soil N availability. In contrast, Díaz et al. (1993) showed that while elevated CO_2 increased the flux of C to soil via roots, an enlarged microbial biomass led to net immobilization of N, thereby limiting the availability of N to plants, creating a negative feedback (Fig. 6.3). In a synthesis of results from forest climate change experiments, however, Zak et al. (2003*a*) found no consistent effects of elevated CO_2 on N mineralization or the availability of N for plant growth.

A key determinant of whether increases in the flux of C to soil lead to immobilization or mineralization of N is the C : N ratio of substrate and the microbial community. If the C : N ratio of the extra substrates being added to soil under elevated CO_2 is low, microbes will become C limited, and will utilize available C, but excrete excess N into the soil environment, thereby increasing N availability to plants (Kaye and Hart 1997). In contrast, if the increased substrate load is of high C : N, microbes will become N limited; hence, they will immobilize N, leading to reduced N availability to plants (Kaye and Hart 1997). In general, elevated atmospheric CO_2 concentrations reduce N concentrations in plant tissues, leading to larger C : N ratios of litter inputs to soil (Curtis and Wang 1998; Coûteaux et al. 1999). In situations where this leads to a switch from C to N limitation of the microbial community, it will therefore potentially lead to microbial immobilization of N, thereby reducing the supply of N to plants and constraining the fertilizing effect of elevated CO_2 on plant growth.

Table 6.2 Effects of elevated atmospheric CO_2 concentrations on soil total inorganic N (N), nitrate (NO_3^-), ammonium (NH_4^+), and N mineralization rate

Study description[a]	Species/community	Group[b]	Inorganic N			Net/gross N mineralization[c]	Reference
			N	NO_3^-	NH_4^+		
370, 570 ppm / 4 yrs / OTC	Spruce (Picea abies) and beech (Fagus sylvatica)	W		→	↓/↑d		(Hagedorn et al. 2002)
350, 525, 700 ppm / 6 yrs / OTC	Ponderosa pine (Pinus ponderosa)	W		0e	0e		(Johnson et al. 2000)
Ambient, +350 ppm / 1 yr / OTC	Scrub oak	W		→	0/↓	↓†	(Hungate et al. 1999)
Ambient, +350 ppm / <1 yr / OTC	Douglas fir	W		→	→		(Janssens et al. 1998)
350, 570 ppm / 4 yrs / OTC	Spruce (P. abies) and beech (Fagus sylvestris)	W	0/↓f			0†	(Hagedorn et al. 2000)
375, 700 ppm / 2 yrs / GH	Yellow birch (B. alleghaniensis)	W			0	0†	(Berntson and Bazzaz 1997)
365, 700 ppm / 2 yrs / GH	Regenerating forest	W				↓†	(Berntson and Bazzaz 1998)
365, 565 ppm / 4 yrs / FACE	Loblolly pine (P. taeda)	W	0			0	(Finzi et al. 2002)
350, 700 ppm / 2 yrs / GH	Sweet chestnut (Castanea sativa)	W				→	(Rouhier et al. 1994)
Ambient, 200 ppm / 2 yrs / FACE	Loblolly pine (P. taeda)	W				0	(Allen et al. 2000)
357, 707 ppm / 3 yrs / OTC	Trembling aspen (P. tremuloides)	W				0†	(Zak et al. 2000)
342, 692 ppm / <1 yr / OTC	Bigtooth aspen (P. grandidentata)	W				↑	(Zak et al. 1993)
367, 715 ppm / 2 yrs / OTC	Trembling aspen (P. tremuloides)	W				0†	(Mikan et al. 2000)
Ambient, 550 ppm / 2 yrs / FACE	Desert ecosystem	W/G/H	→	→	0		(Billings et al. 2002b)
Ambient, 550 ppm / 2 yrs / FACE	Desert ecosystem	W/G/H				0	(Billings et al. 2002a)
Ambient, +200 ppm / <1 yr / GC	Model grassland	G/H				0	(Kampichler et al. 1998)
200–365, 365–550 / 3 yrs / GC	Grassland	G/H				→	(Gill et al. 2002)

Table 6.2 (*Continued*)

Study description[a]	Species/community	Group[b]	Inorganic N	Inorganic NO_3^-	Inorganic NH_4^+	Net/gross N mineralization[c]	Reference
350, 525, 700 ppm / 2 yrs / GC	Grassland	G/H				0	(Newton et al. 1995)
Ambient, 2Xambient / 4 yrs / OTC	Tallgrass prairie	G/H				0	(Rice et al. 1994)
376, 700 ppm / decades / natural CO₂ vent	Grassland	G/H				↑	(Ross et al. 2000)
356, 600 ppm / 5 yrs / OTC	Grassland	G/H		↓	0	0	(Niklaus et al. 2001)
Ambient, +350 ppm / <1 yr / OTC	Sandstone grassland	G/H				↑†	(Hungate et al. 1997a)
	Serpentine grassland	G/H				↑†	
369, 600 ppm / <1 yr / GC	Barley (*H. disticum*)	G	0				(Martin-Olmedo et al. 2002)
350, 700 ppm / <1 yr / GH	Compact brome (*B. madritensis*)	G				↓	(Dhillion et al. 1996)

[a] Atmospheric CO₂ concentrations / experimental duration / design (FACE, free air CO₂ enrichment; OTC, open top chamber; GH, greenhouse; GC, growth cabinet).
[b] Plant species type (W, woody; G, graminoid; H, herbaceous).
[c] Net N mineralization, except† which denotes gross N mineralization.
[d] Decreased in acidic soil, increased in calcareous soil.
[e] Water soluble N.
[f] No change in acidic loam, decreased in the calcareous soil.
Notes: ↑, 0, ↓ represent increased, no change and decreased, respectively.

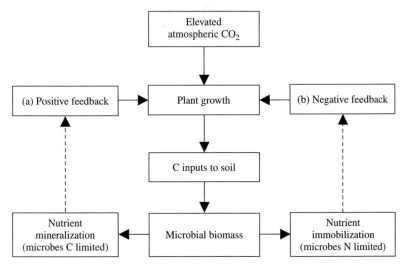

Fig. 6.3 Schematic diagram of the potential effects of elevated atmospheric CO_2 on soil C inputs and microbial biomass, leading to either (a) positive or (b) negative feedback on nutrient availability and plant growth. Positive feedback occurs when the increased supply of root-derived C to soil leads to increased microbial biomass and N mineralization, and uptake of N by plants. Negative feedback occurs when elevated CO_2 increases the flux of C to soil via roots, but an enlarged microbial biomass leads to net immobilization of N, thereby limiting the availability of N to plants. (Zak et al. 1993 and Díaz et al. 1993.)

Another route by which elevated CO_2 may impact on soil biological processes is by intensifying competition between soil microbes and plants for N. This was shown to be the case by Hu et al. (2001). These authors showed that continuous exposure of an annual grassland to elevated CO_2 for five years led to increased plant N uptake, but reduced soil N availability and microbial respiration per unit biomass, thereby exacerbating N constraints on microbes. This finding was taken to suggest that increased CO_2 could alter plant–microbial competition for N, thereby suppressing microbial decomposition and increasing ecosystem C accumulation (Hu et al. 2001). The important message here is that effects of elevated CO_2 on ecosystem performance are likely to relate strongly to the factors that regulate partitioning of nutrients between plants and soil microbes, which, as discussed in Chapter 4, are highly context dependent, varying with plant identity, soil fertility, climate, and a range of other ecosystem parameters.

6.2.3 Influence of soil N availability on ecosystem responses to elevated CO_2

There is mounting evidence that soil nutrient availability is a key determinant of the long-term response of ecosystems to elevated CO_2. The picture emerging is that while elevated CO_2 typically increases plant growth in the short term, longer-term increases will only occur if there is a sustained increase in nutrient use efficiency or there is a continuing supply of N (Finzi et al. 2002). This increase in N can be met by: (1) N reallocation within the plant, (2) increased mineralization of soil and litter by soil biota, or (3) additional supply of N from external sources, such as fertilizers or atmospheric N pollution (Beedlow et al. 2004). A number of modelling studies support the concept of N limitation, showing that the observed increase and subsequent slowing of plant growth in response to elevated CO_2 is a consequence of nutrient limitation (e.g. Pan et al. 1998). Models also predict that soil N availability is a primary constraint on the ability of forests to sequester C. For example, McKane et al. (1997*b*) applied a biogeochemical model to forest stands in the Pacific Northwest, USA, and predicted that over a 100 year period elevated CO_2 and temperature would raise total ecosystem C storage by less than 10% in N-poor sites compared to 25% in N-rich sites. Similarly, vegetation models predict much lower C sequestration in response to climate change when N is limiting than when N is replete (Pan et al. 1998; Hungate et al. 2003).

If elevated CO_2 leads to increased storage of C, this places an additional demand on the availability of N in soil. This is because the storage of C requires the removal of N from the actively cycling pool and sequestering it along with C in wood, leaves, litter, or soil organic matter; this creates a continued demand for N and other nutrients (Beedlow et al. 2004). This decline in N availability under elevated CO_2 is central to the progressive N limitation (PNL) concept of Lou et al. (2004). The central idea of PNL is that increased C flow into an ecosystem at elevated CO_2 is: (1) used for the production of plant biomass; (2) stored in soil organic matter; and (3) returned to the atmosphere through autotrophic and heterotrophic respiration. The additional growth of long-lived plant biomass (e.g. wood in forests) and increased C storage in soil cause N to be sequestered in soil organic matter, progressively decreasing soil mineral N availability for plant use in the long-term. Therefore, in the absence of new N inputs from fertilization or N pollution, or decreases in N loss, the availability of N declines over time at elevated CO_2 in comparison with its availability at ambient CO_2 levels (Lou et al. 2004). Overall, it is proposed that PNL will develop only if elevated CO_2 causes long-lived plant biomass and soil organic matter to accumulate, sequestering substantial amounts of C and N in long-term pools. By contrast, if elevated CO_2 does not stimulate enough biomass growth and C accumulation in soil, N sequestration in long-lived biomass and soil organic matter will not be substantial enough to affect N availability; in this case, PNL may not develop (Fig. 6.4).

6.2.4 Elevated CO_2 and plant community composition

As already noted, elevated atmospheric CO_2 concentrations generally increase the C : N ratio of plant tissue (Curtis and Wang 1998; Coûteaux

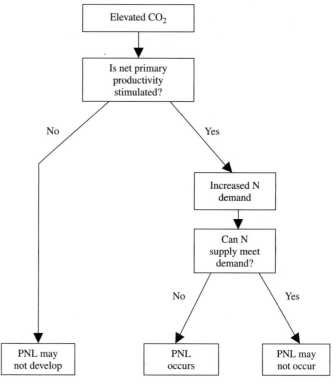

Fig. 6.4   Schematic diagram of the progressive N limitation (PNL) framework of Lou et al. (2004) for identifying patterns of interactions between C and N under elevated CO_2. If elevated CO_2 does not stimulate enough biomass growth and C sequestration in soil, N sequestration in long-lived plant biomass and soil organic matter will not be substantial enough to affect soil N availability, so PNL may not occur. If stimulation of C sequestration under elevated CO_2 is high, N demand will build up to balance the C influx. If this extra N demand can be met through internal N cycling or inputs of N from outside, PNL may not occur. However, if N supplies cannot satisfy the N demand, PNL may occur. (Lou et al. 2004.)

et al. 1999), thereby altering the quality and decomposability of litter inputs to soil. Another route by which increasing CO_2 could influence the quality of litter inputs to soil is through CO_2-driven shifts in plant community composition. Few studies have explicitly tested this idea, but it is well documented that increasing CO_2 can lead to shifts in the functional composition of grassland communities. For example, it has been shown by Niklaus et al. (2001) that elevated CO_2 can affect plant succession in temperate grassland, and Körner et al. (1997) showed that while the overall biomass of a calcareous grassland was unaffected by elevated CO_2, there was a shift away from the dominant sedge *Carex curvula* to neighbouring subordinate species. In another study of calcareous grassland, Hanley et al. (2004) showed that elevated CO_2 increased total community biomass when receiving additional nutrients, and that this was associated with an increase and decrease in the biomass of forbs and grasses, respectively. Legumes were also found to increase in biomass under elevated CO_2, but only in the absence of additional nutrients (Hanley et al. 2004). The finding that elevated CO_2 favours legumes over grasses is quite common (e.g. Newton et al. 1995; Ross et al. 2004), and potentially has great significance for the N economy of grassland ecosystems under climate change.

CO_2-driven changes in plant community composition and diversity could affect soil biological properties in a number of ways. First, resulting changes in the composition and diversity of litter inputs to soil could greatly influence the rate of decomposition and hence availability of nutrients in soil. As discussed in Chapter 4, it is well known that decomposition rate changes when different leaf species are mixed (e.g. Chapman et al. 1988; Blair et al. 1990) and this process is also affected by variations in the diversity and evenness of litter inputs, albeit in an unpredictable way (Wardle et al. 1997*b*; Bardgett and Shine 1999; Hector et al. 2000; King et al. 2002). Second, they will alter the range of plant functional traits that affect C acquisition by photosynthesis, thereby potentially influencing the amount of C returned to soil both as plant litter and as root exudates. Extremely little is known about how variations in plant diversity affect C acquisition in a high CO_2 world, but a study of grassland showed that the positive effect of elevated CO_2 on plant productivity was less marked in species-poor than in species-rich grassland communities (Reich et al. 2001*a*), in part owing to a reduced range of plant functional traits that affect C acquisition and cycling (Reich et al. 2001*b*). This finding was taken to suggest that future reductions in botanical diversity of ecosystems might reduce their capacity to capture additional C in a high CO_2 world (Reich et al. 2001*a*), thereby also having consequences for C supply to soil. It should be noted that others have not found strong relationships between plant diversity and C assimilation (e.g. Stocker et al. 1999), and, to date, no studies have tested explicitly the hypothesis that reductions in plant diversity will impair the ability of ecosystems to sequester C under elevated CO_2. A final route by which shifts in the composition of vegetation could affect soil biological properties is via an increase in the influence of legumes on soil N cycling. As noted in Chapter 4, legumes greatly increase inputs of N to soil and thereby increase N availability. Therefore, an increase in the occurrence of legumes or other N fixers within the plant community could reduce N limitation, thereby removing a major constraint on plant and microbial responses to elevated CO_2.

6.2.5 Effects of elevated temperature

Relatively few studies have looked at the potential indirect effect of elevated atmospheric CO_2 on soil biological properties, notably an increase in mean atmospheric temperature of the order of some 1–3 °C (Houghton et al. 1995). It has been suggested that such small changes in temperature will have relatively little effect on soil biota due to their wide temperature optima (Tinker and Ineson 1990), and this view is generally supported by a number of literature syntheses that report only weak relationships between mean temperature and densities of major faunal groups (Petersen and Luxton 1982; Didden 1993; Wardle 2002) and also microbial biomass (Wardle 1992). Also, the few experimental studies that have been done on warming show a general lack of effects of elevated temperature on below-ground communities and processes (Hodkinson et al. 1996; Kandeler et al. 1998; Webb et al. 1998; Bardgett et al. 1999*b*), again suggesting limited effects of small rises in temperature on soil biota. Despite the general insensitivity of soil biota to small changes in temperature, there are situations where warming could have significant effects on soil biological processes, especially under extreme climates where low temperatures and hydrology act as primary determinants of ecosystem structure. This section will examine the potential effects of climate change on soil and ecosystem properties of two such ecosystem types: peatland and arctic ecosystems, which are both thought to be especially prone to climate warming.

Effects of warming on northern peatland ecosystems

A key concern is that increases in atmospheric temperature could enhance the turnover of soil organic C (Jenkinson et al. 1991; Trumbore et al. 1996), leading to the release of C in gaseous form and as DOC in drainage waters (Raich and Schlesinger 1992; Freeman et al. 2001). Such a release of C is expected to be greatest from extensive peatlands of wetland, tundra, and Boreal zones, which represent a vast store of terrestrial C, currently estimated to be some one-third of global C stock (Post et al. 1982; Gorham 1991) (Fig. 6.5). The concern here is that climate warming will destabilize these stores, releasing C in either gaseous or aqueous forms to the surrounding environment (Freeman et al. 2001). These concerns appear to be supported by observations of rapidly rising DOC concentrations in aquatic ecosystems that receive drainage waters from peatlands in the United Kingdom (Freeman et al. 2001; Worral et al. 2003, 2004) and in North America (Schindler et al. 1997). The cause of such increases in aquatic DOC has been the subject of much debate, and several hypotheses have been proposed. One explanation is that long-term increases in temperature have increased rates of peat decomposition owing to increased activity of decomposer organisms (Freeman et al. 2001; Worral et al. 2004). This view is supported to some extent by studies of decomposition processes in peatlands of northern England. Here, transplanting of peat soils to lower altitudes was used to simulate climate warming, and was found to increase the density of enchytraeid worms, the dominant fauna of these peat soils, leading to enhanced decomposition and DOC concentrations

Fig. 6.5   A section of blanket peat at Moor House National Nature Reserve, northern England. These peatlands are of special significance because they represent a significant (ca., 30%) store of global terrestrial C. The concern here is that climate warming will destabilize these stores, releasing C into waters as DOC. (Image by Richard Bardgett.)

in soil leachates (Briones et al. 1997, 1998). Similarly, in a field study, Cole et al. (2002a) showed that field densities of these animals are strongly related to temperature and proposed that soil warming could potentially increase enchytraeid population densities, especially early in the growing season, leading to an overall increase in the release of DOC from the blanket peat ecosystems (Fig. 6.6).

The general consensus, however, is that increased decomposition resulting from long-term warming accounts for only a small portion of the total DOC release from peatland (Freeman et al. 2001, 2004; Worral et al. 2004). This is because laboratory and field studies of peat soils show little relationship between temperature and DOC release (Hudson et al. 2003; Worral et al. 2004), suggesting that other processes are at play. An alternative mechanism commonly proposed is that the increased incidence of summer drought leads to enhanced decomposition under warmer, drier conditions, and the mobilization of soluble C following the return to wetter conditions (Tipping et al. 1999; Worral et al. 2003, 2004). This view, however, has recently been challenged by Freeman et al. (2004). These authors subjected a chain of Welsh peatlands to repeated, simulated drought and found no effect on DOC release. They did find, however, that elevated CO_2 significantly increased DOC release from peatlands in the long term, an effect that was especially pronounced in more productive sites where plant productivity was less constrained by nutrient limitation (Fig. 6.7). This was attributed to an increase in plant productivity and root

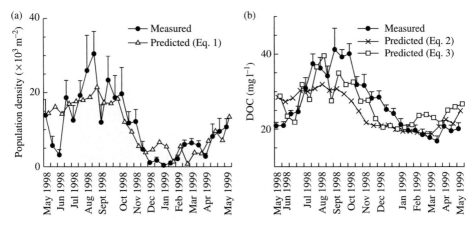

Fig. 6.6 (a) Population density of enchytraeids in blanket bog at Moor House, estimated by wet funnel extraction and enumeration (measured), and predicted from the regression relationship between population density and soil temperature (predicted). Population density of enchytraeids was measured by wet extraction and is expressed as the mean of the total number of enchytraeids extracted from the entire 0–15 cm depth, error bars represent SE ($n = 5$). (b) DOC concentration at 10 cm depth at same site, as well as estimated values from the relationships between soil temperature (Eq. 2) and from the enchytraeid population data (Eq. 3). The gradient of Eq. 3 is the production of soil DOC at the study site for an associated increase in the total enchytraeid population of 1000 m^{-2}. The present mean total population density of enchytraeids at the site is 10.7×10^3 m^{-2}, whereas the predicted mean total population density of enchytraeids at the site for a 2.5 °C increase in temperatures was 15.3×10^3 m^{-2}. This represents an overall increase of 4.6×10^3 m^{-2} in the enchytraeid populations, which would cause an increase of ca. 2 mg l^{-1} in soil solution DOC at the surface. This corresponds to an overall increase in DOC concentrations at 10 cm depth to 28.5 mg l^{-1}, representing an increase of 8% for a 2.5 °C increase in mean monthly air temperatures. (Data from Cole et al. 2002a.)

exudation of recently assimilated CO_2 under elevated CO_2, which was confirmed using ^{13}C tracer studies of peat monoliths. These studies showed that the proportion of DOC in soil solution derived from recently assimilated CO_2 was 10 times greater under elevated CO_2 than in the ambient control. Overall, these findings were taken to suggest that DOC release from peatland is far more sensitive to atmospheric CO_2 than to warming or the hydrological changes associated with droughts, and that the primary mechanism for this is enhanced primary productivity and root exudation (Freeman et al. 2004).

Effects of warming in high arctic ecosystems

Low temperatures are one of the primary determinants of high arctic ecosystems and thus increases in temperature could have profound impacts on their structure and function (Fig. 6.8). Climate change scenarios predict that before the end of the current century, summer warming over the arctic land mass could be as high as 4–8 °C, and might be even greater in the winter (Oechel et al. 1997). Even over the last few decades, mean annual

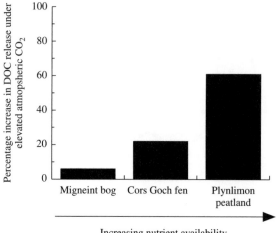

Fig. 6.7   Effects of elevated CO_2 on DOC release from a range of peatland sites in Wales, expressed as % increase in DOC release under elevated CO_2 (compared to ambient treatment). Under elevated CO_2, DOC release from all three sites of differing inherent fertility increased significantly, but the greatest increase was found in the site with the highest nutrient status, that is, Plynlimon. Migneint bog was nutrient poor, receiving nutrients solely from rainfall; Cors Goch fen gains more nutrients as a consequence of influxes from surrounding soils and ground-waters, and; Plynimon peat gains even more nutrients by intercepting nutrient-laden waters from adjacent terrestrial systems. (Data from Freeman et al. 2004.)

Fig. 6.8   Low temperatures are one of the primary determinants of high arctic ecosystems and thus increases in temperature could have profound impacts on their structure and function. (Image by Richard Bardgett.)

temperatures in the Arctic have risen by some 2–4 °C (Lachenbruch and Marshall 1986; Oechel et al. 1993), largely due to an increase in winter and spring temperatures (Chapman and Walsh 1993). As in peatland of Boreal zones, the main concern here is the potential for warming to stimulate decomposition and C loss from seasonally thawed soils, which presently contain some 12% of the global C store, even though they only occupy about 5% of the total land mass (Billings 1987; Schlesinger 1991).

During the growing season, rates of N mineralization in the Arctic are typically very low, or even negative (Nadelhoffer et al. 1991; Schmidt et al. 1999*a*; Jonasson et al. 1999*a*), and it is generally thought that rising temperature will increase rates of N mineralization, leading to increased nutrient supply to plants (Jonasson et al. 1999*b*; Ruess et al. 1999). This in turn is predicted to change plant community composition in favour of species with greater nutrient uptake rates, such as graminoids and deciduous shrubs, at the expense of plants with lower uptake rates and growth rates, such as mosses and evergreen shrubs (Chapin et al. 1996*a*; Press et al. 1998). However, it is also possible that under warming, soil microbes will act as a strong nutrient sink, out-competing plants for any available nutrients (Jonasson et al. 1999*b*). It is well documented that microbes compete effectively with plants for nutrients in the Arctic (Schimel and Chapin 1996), and that they act as a strong nutrient sink in these systems, holding as much as 7% of the total N pool and 35% of the total P pool (Jonasson et al. 1996, 1999*a*). Therefore, it is possible that increases in plant production due to warming will not be paralleled by increases in nutrient supply to plants, thereby leading to reduced foliar nutrient contents, and reduced decomposability of litter. This was shown to be the case by Welker et al. (1997) who reported that warming increased growth of *Dryas octopetala*, a dominant arctic plant, but reduced its foliar N content (and increased its C : N ratio), which would presumably reduce decomposition, lowering nutrient availability, and possibly lowering plant N levels and growth in subsequent years (Fig. 6.9).

Changes in the composition of vegetation resulting from climate change could also have consequences for soil biological properties in the Arctic, affecting the quality and quantity of resource inputs to soil. For example, Hobbie et al. (1999) conducted field experiments in arctic tundra, and found that elevated temperature favoured sedges and woody deciduous shrubs, such as *Betula nana*, over *Eriophorum* and non-vascular plants. The consequences of this change in vegetation for decomposition and nutrient cycling were not studied, but those species that were promoted by warming, such as deciduous shrubs, produce large amounts of low quality, woody litter that decomposes slowly, potentially leading to C accumulation on the soil surface (Hobbie 1996) and effects on the hydrology of soil. Modelling studies also provide evidence of large shifts in vegetation that could result from climate warming. For example, Starfield and Chapin (1996) predicted that a rise in temperature of 3 °C would be enough to

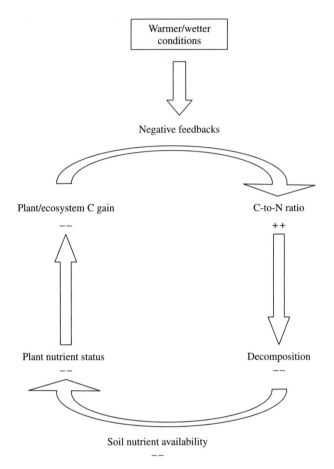

Fig. 6.9   Conceptual model of the direct impacts of climate warming on *Dryas octopetala* plants, leading to negative feedbacks on arctic ecosystem properties. (Welker et al. 1997.)

induce some arctic tundra communities to convert to coniferous forest within some 150 years, and Epstein et al. (2000) predicted, using simulation models, that warming will increase in shrub abundance in arctic tundra.

Effects of climate warming in the Arctic are not confined to processes that occur over the short growing season. An important consequence of climate change in high latitude and altitude regions is likely to be a change in winter precipitation and snow cover (Maxwell 1992). Snow cover is an important determinant of the arctic and alpine ecosystem, affecting growing season length, soil moisture, and temperature (Walker et al. 1999). Snow is also a critical regulator of ecosystem processes due to its role as an insulator of soil (Brooks et al. 1995), promoting the activity of soil microarthropods (Addington and Seastedt 1999) and microbes (Brooks et al. 1996), and increasing rates of nutrient mineralization (Schimel et al. 2004). It has been predicted that a reduction in snow cover in arctic regions will potentially result in colder soil temperatures, more extensive freezing, and an increase in the frequency of freeze–thaw cycles over the winter months (Edwards and Cresser 1992). Not much is known about the consequences of reduced snow cover, but a number of studies show changes in winter snow depth can greatly impact on microbial respiration and N mineralization (Brooks et al. 1995; Schimel et al. 2004), and that increased freeze–thaw activity can lead to major declines in microbial biomass and consequent release of nutrients held within microbial cells (Schimel and Clein 1996; Larsen et al. 2002). The occurrence of surface ice layers—which are predicted to occur more frequently as a consequence of climate change—have also been shown to have strong negative effects on soil microarthropods, reducing their numbers by some 50% (Coulson et al. 2000). As discussed in Chapter 4, winter soil processes are increasingly being recognized as being of great importance for nutrient and C cycling in cold regions. Hence, such effects of climate change on snow cover and freezing could have far reaching consequences for the structure and functioning of arctic ecosystems.

The overall effects of climate warming on net ecosystem C flux in the Arctic will depend on the balance between C gains through changes in plant productivity and losses from increased microbial activity (Shaver et al. 2000). It will also depend greatly on the effects of warming on soil hydrological conditions and especially on permafrost, a key determinant of plant growth and decomposition processes in the Arctic (Oechel et al. 1997). Most estimates indicate that under climate warming the Arctic will switch from being a C sink to a C source (Oechel et al. 1997; Welker et al. 1999), and several studies reveal that many arctic ecosystems are already acting as CO_2 sources to the atmosphere as a result of climate change (Oechel et al. 1997). Given the large C stores in arctic soils, this loss of C may represent a significant and positive feedback to atmospheric CO_2 concentration and concomitant global change.

6.2.6   Soil carbon sequestration

Carbon sequestration implies transferring atmospheric CO_2 into long-lived soil organic C pools where it is stored securely so it is not immediately reemitted (Lal 2004) (Box 6.1). This issue is high on the political and scientific agenda, largely owing to the interest in the extent to which natural and managed terrestrial ecosystems can act as a sink for C and how this could help to mitigate human-induced increases in atmospheric CO_2; indeed, terrestrial sinks have been included in the Kyoto Protocol to meet targets for reduced greenhouse gas emissions. Whether natural ecosystems act as a C source or sink under climate change is very uncertain, since it depends on so many factors that could potentially affect the equilibrium soil C status, depending on how the balance between C loss through decomposition and C gain from primary productivity are affected

(Fig. 6.10). As discussed in the preceding sections, this is likely to be highly context dependent, varying with a suite of factors such as soil fertility (i.e. nutrient availability), hydrology, vegetation type, and susceptibility to extreme weather events (e.g. freeze–thaw and wildfire). While some models predict that increases in primary productivity resulting from elevated CO_2 will increase soil C sequestration (e.g. Post et al. 1992), there is not much experimental evidence to support this. As discussed previously, nutrient limitation is likely to constrain plant responses to rising CO_2, thereby limiting the ability of natural ecosystems to sequester C. Furthermore, CO_2-driven increases in the allocation of labile C below-ground can stimulate fine root production in forests, which in turn increases soil microbial activity and respiration, thereby balancing the increased C input

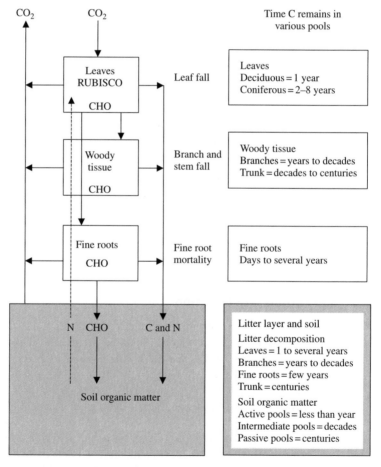

Fig. 6.10 Schematic diagram of C uptake, release, and retention in forest ecosytems. CHO represents the movement of photosynthates between plants, soils, and the atmosphere. (Redrawn with permission from the Ecological Society of America; Beedlow et al. 2004.)

to soil, with no net effect on long-term soil C storage (Schlesinger and Lichter 2001). Increased allocation of C below-ground may also stimulate microbial activity to such an extent that it leads to enhanced turnover of native soil organic C—the so called priming effect—increasing the flux of CO_2 from soil, thereby reducing soil C storage (Körner and Arnone 1992; Hungate et al. 1997*b*; Fontaine et al. 2004). It would seem unlikely, therefore, that rising CO_2 will cause a sustained increase in C sequestration in most natural ecosystems (Beedlow et al. 2004).

Managing soils for C sequestration is perhaps most relevant to the world's agricultural and degraded soils, which offer great potential to act as C sinks under judicious management (Lal 2004; Smith 2004). This is because agricultural soils have been subject to tremendous C loss, from initial conversion from natural lands and subsequent continuous cultivation and grazing which progressively degrades soil C stocks. A range of strategies have been proposed for long-term enhancement of the C pool in agricultural and degraded soils, such as the adoption of no-tillage agriculture, the use of cover crops, and the adoption of low intensity grazing systems (Lal 2004). A good example of how changes in management can promote soil C storage comes from France, where large areas of grassland have been converted to arable land since the 1970s, leading to significant declines in soil C status. It has been predicted that the restoration of half of this arable

Box 6.1 Soil organic matter pools

Soil organic matter is located within more-or-less 'discrete' pools in the soil. The most active, rapid cycling pool of C consists of microbial biomass and easily decomposable plant residues; this material cycles rapidly, and hence makes up only a small fraction of the total soil organic matter in soil (3–5%). The other major pool of C in soil is that which is either physically or biochemically protected from decomposition; this soil C is referred to as old C in that it has a much longer residence time in soil, ranging from tens-to-thousands of years. This C may be physically protected from decomposition through occlusion by clay minerals and encapsulation within soil aggregates, which forms a physical barrier between microbes and their substrates. Alternatively, soil C may be biochemically protected from decomposition through chemical complexing processes between substrates, such as lignins and polyphenols, and soil particles. With respect to soil C storage, that which is located in soil macroaggregates (>200 μm) typically has a faster turnover rate, and hence shorter residence time in soil, than that within microaggregates (50–200 μm) (Jones and Donnelly 2004). The most stable C in soil is that which is either complexed with clay minerals or biochemically protected.

land back to permanent grassland (~90,000 ha) would eventually lead to an estimated increase of 16 Mt of soil organic C, equivalent to 10% of the annual CO_2 emissions from fossil fuels in France (Soussana et al. 2004). Other changes in management that have been shown to greatly increase soil C accumulation include the widescale adoption of no-tillage agriculture. In Brazil, for example, the ongoing conversion to no-tillage agriculture has led to the accumulation of some 9 Mt C yr^{-1} in soil (Cerri et al. 2004); in the United States, the adoption of no-tillage agriculture is thought to potentially increase soil C storage by some 337 kg C ha^{-1} yr^{-1} (West and Marland 2002), and; in tropical Nigeria, the adoption of no-tillage maize has been shown to increase soil C by some 0.17 t C ha^{-1} yr^{-1} over conventional tillage operations (Lal 1997).

In considering beneficial effects of land use change on C sequestration it is important to consider associated impacts on the net emission of other potent greenhouse gasses, such as methane (CH_4) and nitrous oxide (N_2O). This is especially the case for no-tillage agriculture. While promoting soil C storage, this practice also potentially increases emissions of N_2O to the atmosphere, owing to increased denitrification in compacted, low porosity soils (Smith and Conen 2004). Such increases in the emission of N_2O will affect the net greenhouse gas balance of systems, having the potential to negate benefits gained from increased C storage. This has been shown to be the case by Six et al. (2004). These authors compiled global data on soil-derived greenhouse gas emissions (i.e. CO_2, N_2O, and CH_4) in no-tillage and conventional tillage systems and found that newly converted no-tillage systems had a greater global warming potential than did conventional systems—largely due to increased N_2O emission—and that no-tillage only reduced global warming potential in the long-term (>20 years) in humid climates. In dry areas, reductions in global warming potential declined only after 20 years, but with a high degree of uncertainty. In view of this, it has been argued that the promotion of no-tillage and other farming practices for C sequestration is perhaps naïve since its effect on the net greenhouse balance is highly variable, complex, and time dependant (Six et al. 2004; Smith and Conen 2004).

6.3   Atmospheric N deposition

Post-industrial human activities have increased substantially the release of reactive N into the global environment, greatly affecting the global N cycle. This increase in N has resulted from a range of human activities, including the fixing of N_2 for fertilizers, the burning of fossil fuels, increasing use of legumes in agriculture, and mobilization of N from long-term biological stores, such as forests and wetlands. Such alteration of the N cycle has multiple consequences, including increasing concentrations of the greenhouse gas N_2O in the atmosphere, and a substantial increase in the flux of nitrous oxide (NO_x) and ammonia (NH_3) gas. Enhanced emission of reactive N to the atmosphere has in turn led to increased deposition of N on land and oceans, having major consequences for recipient ecosystems. This section will examine the effects of this enhanced N deposition for soil biota. It will also consider the involvement of soil organisms in the retention and loss of N from polluted soils.

6.3.1 Effects of N enrichment on plant and soil biological communities, and ecosystem C turnover

Most natural and semi-natural plant communities are N limited, so the addition of N will substantially alter their structure and productivity, favouring productive, fast-growing species that are best able to use this added resource. This has been shown to be the case in several situations. For example, studies from moss-dominated montane heath in the Highlands of Scotland show that sedges and grasses increase in abundance under conditions of enhanced N deposition (Pearce and Van der Wal 2002; Van der Wal et al. 2003) (Fig. 4.14). This has, in turn, been shown to lead to the demise of mosses, owing to both direct negative effects of N on moss growth (Woodin and Lee 1987; Carroll et al. 2000; Gordon et al. 2001; Pearce et al. 2003) and indirect effects of reduced light availability on low stature mosses, resulting from increased canopy closure by vascular plants (Van der Wal et al. 2005). Arctic communities also respond to N addition, often leading to an increase in the abundance of woody deciduous shrubs, such as *Betula nana* (Bret-Harte et al. 2002), or increases in biomass of graminoids (Press et al. 1998; Shaver et al. 1998; Gough et al. 2002), depending on context. Studies of temperate grasslands also reveal dramatic effects of N on plant community structure. For example, in The Netherlands, high levels of N deposition have led to the conversion of infertile, species-rich heathlands to species-poor grassland and forest of reduced conservation value (Aerts and Berendse 1988). It is also well documented that N fertilization of grasslands promotes the growth of competitive, fast-growing grasses at the expense of herbs and legumes, thereby reducing plant diversity (Van Dam et al. 1986; Mountford et al. 1993; Wedin and Tilman 1996). The implications of this for plant species diversity in grassland are clear from the study of Stevens et al. (2004). These authors reported that long-term, chronic N deposition across Britain (ranging from 5-to-35 kg N ha^{-1} yr^{-1}) has significantly reduced plant species richness in grassland, and that species richness has declined as a linear function of the rate of N deposition, with a reduction of 1 species per 4 m^2 for every 2.5 kg N deposited ha^{-1} yr^{-1}. Furthermore, they reported that at the mean chronic N deposition rate of central Europe (17 kg N ha^{-1} yr^{-1}), there has been a 23% reduction in species richness compared with grasslands receiving the lowest levels of N deposition.

The above, relatively predictable changes in vegetation resulting from N enrichment should potentially increase both the quantity and quality of litter inputs to soil, thereby promoting soil biological activity and nutrient cycling. While positive effects of N enrichment on soil microbes have been detected (e.g. Hart and Stark 1997; Scheu and Schafer 1998) there are many situations where it has had neutral or even negative effects on soil microbes and their activity. For example, Schmidt et al. (2004) found that N fertilization of high elevation tundra soils in Colorado depressed microbial biomass and activity, especially during the summer, and Johnson et al. (1998) found that seven years of simulated N deposition to calcareous grassland in the United Kingdom significantly depressed soil microbial biomass, but the same treatment applied to acid grassland had no effect on this measure. Similarly, N enrichment had negative effects on soil respiration in forest ecosystems (e.g. Söderström et al. 1983; Burton et al. 2004), but positive and neutral effects have also been reported (Nohrstedt and Börjesson 1998; Kane et al. 2003).

While effects of N deposition on microbial biomass are highly context dependent, its influence on the structure of microbial communities appears to be a bit more predictable. In many ecosystems, such as hardwood forest and grassland, a striking effect of N enrichment is a depression in the biomass of decomposer fungi, leading to a shift towards a bacterial-dominated food web (Bardgett et al. 1993a; Bardgett and McAlister 1999; Donnison et al. 2000a,b; Grayston et al. 2001a; Frey et al. 2004). The abundance and diversity of EM (Lilleskov et al. 2001; Erland and Taylor 2002; Frey et al. 2004) and AM fungi (Donnison et al 2000b; Egerton-Warburton and Allen 2000) is also often reduced by N pollution and N fertilization, but effects on AM fungi appear to vary in soils of different ambient soil fertility. As shown by Johnson et al. (2003c), N fertilization decreases allocation to AM structures at sites with ample P (i.e. low N : P ratio of soil), whereas it increases allocation to AM fungal structures when P is limiting (i.e. high N : P ratio). This positive effect of N enrichment on AM fungi in P-limited soils is likely to occur because N fertilization exacerbates P limitation and further increases the roles of mycorrhizae for plant P uptake. In contrast, N fertilization of P enriched soils (i.e. low N : P) eliminates N limitation causing plants to reduce investment below-ground (Johnson et al. 2003c). N enrichment also has strong impacts on higher trophic levels, but as with microbial biomass, these effects are typically very variable. This variability is thought to reflect the varying degree to which fauna are regulated by 'bottom-up' (i.e. resource supply) or 'top-down' (i.e. predation) forces, in that those organisms regulated primarily by bottom-up forces are more likely to be affected by N enrichment than those regulated by 'top-down' forces (Wardle 2002).

One idea that has been proposed to explain the variable effect of N deposition on the activities of soil microbial communities is that ecosystems with more recalcitrant litter tend to respond negatively to N deposition, whereas ecosystems with rapidly cycling litter respond positively (Waldrop et al. 2004). This idea is based on the general finding that high levels of inorganic N enrichment often stimulate degradation of labile, high cellulose litter, whereas they often suppress breakdown of recalcitrant, high lignin litter (Fog 1988). These very different responses are thought to reflect differential effects of N enrichment on the activities of extracellular enzymes involved in cellulose and lignin degradation, in that the ability of microbes to synthesize cellulases is typically enhanced by N enrichment, whereas the synthesis of ligninolytic enzymes by white rot fungi is generally suppressed (Hammel 1997; Carreiro et al. 2000; Frey et al. 2004). Therefore, ecosystems with more recalcitrant litter contain more white rot fungi that respond negatively to N deposition, whereas ecosystems with rapidly cycling litter have different microbial communities which respond positively to N deposition (Waldrop et al. 2004). This framework also accords with the common finding that N enrichment can have variable effects on rates of C turnover during different stages of decomposition. In a literature synthesis, for example, Berg and Matzner (1997) observed that N enrichment most often stimulated initial decomposition of litter, but it suppressed humus decay in later stages. Similarly, Neff et al. (2002) found that while N fertilization of an alpine meadow promoted decomposition of light soil organic matter fractions, it stabilized soil organic matter in older fractions. Such findings have clear implications for soil C sequestration, suggesting that N enrichment could influence C storage in soils by stabilizing stores of old and recalcitrant litter.

How N deposition influences the capability of soil to sequester C in a warming climate is a subject that is high on the political and scientific agenda, and represents a major research challenge for the future. There is not much data available on the effects of N fertilization and ecosystem C storage under climate change, but it has been suggested that ecosystems that are subject to high N loads could act as localized sinks for CO_2, as a result of increases in primary productivity and therefore C sequestration (Lloyd 1999). Studies of tundra ecosystems also indicate that increases in nutrient availability, resulting from warming of soils or N pollution, could increase the capacity of these ecosystems to store C (Gorham 1991; Hobbie 1996; McKane et al. 1997a). This is because more C will be stored above-ground as a result of a stimulation of plant productivity and a shift in plant species composition from slow-growing species to more productive deciduous shrubs, such as *Betula nana*, that accumulate C in long-lived woody tissue, which decomposes more slowly (Goulden et al. 1998; Bret-Harte et al. 2002). However, the view that fertilization would increase total C storage in these ecosystems has been challenged by Mack et al. (2004), who showed that long-term additions of N to Alaskan tundra actually reduced ecosystem C storage owing to losses of C from deep soil layers, which more than offset increased C storage in plant biomass and litter. The mechanism for this appears to be a stimulation of decomposition of old plant litter and soil organic C in deep soil layers, leading to loss of C via mineralization and leaching of dissolved organic C. Overall, these findings suggest that combined effects of warming and N deposition in these tundra ecosystems, which act as a major store of global C, may further amplify C release from soils, causing a net loss of ecosystem C and a positive feedback to climate warming.

6.3.2 The retention and export of pollutant N

Understanding the fate of anthropogenic N additions in soils, and especially the factors that regulate its retention and export, has considerable implications for the functioning of recipient terrestrial ecosystems and the environment beyond. In particular, changes in soil N retention that accompany increasing N deposition determine the export of N to adjacent aquatic ecosystems, where it can have substantial, harmful effects on water quality owing to eutrophication and acidification. Also, the loss of gaseous N to the atmosphere has implications for air quality, since it acts as one of the more potent greenhouse gases alongside CO_2. In general, little is known about the capacity of soils to act as N sinks, or the factors that regulate their ability to do this. However, since most ecosystems are N-limited, it is generally thought that plants will act as sinks for large amounts of anthropogenic N and that this may slow down, or even halt, these deleterious effects of N deposition (Aber et al. 1989). It also appears as if some soils have a very high capacity to accumulate and retain N in stable organic matter pools, thereby reducing N export and harmful effects of pollution. This has been shown to be the case in semi-natural grasslands in Britain, for example, where significant amounts of N added over several years have accumulated in soil organic matter, thereby reducing the effects of N pollution on these systems and also the export of N to the environment beyond (Phoenix et al. 2004).

It has been proposed that one of the most likely routes of N storage in soils is the rapid immobilization of N inputs by soil microbes and its subsequent transfer to plants and/or stable, non-microbial organic matter pools (Zogg et al. 2000). A number of studies support this notion. For example, a field study of alpine tundra showed that soil microbes act as a large, albeit short-term, sink for simulated anthropogenic N inputs, especially towards the end of the growing season (Fisk and Schmidt 1996). Similarly, ^{15}N labelling studies in temperate hardwood forests show that soil microbes act as a large initial sink for inputs of ^{15}N–NO_3^-, and that subsequent microbial turnover and cell death leads to the transfer of this microbial N to plants and/or stable organic matter pools where it is retained (Zogg et al. 2000) (Fig. 6.11). Immobilization into the microbial biomass also represents a significant and immediate sink for added N in semi-natural temperate grasslands (Jackson et al. 1989; Bardgett et al. 2003) (Fig. 6.12), and its subsequent transfer to non-microbial, recalcitrant organic matter pools represents an important long-term N sink in these ecosystems (Phoenix et al. 2004). These studies all point to the importance of microbial immobilization for N retention in polluted ecosystems and suggest that there is need for

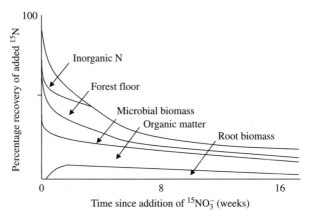

Fig. 6.11 Schematic diagram of total recovery of added $^{15}NO_3^-$ in inorganic-N, microbial, soil organic matter, and root biomass pools, throughout a four-month growing season in a northern hardwood forest. (Adapted from Zogg et al. 2000.)

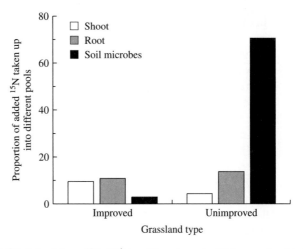

Fig. 6.12 Uptake of ^{15}N derived from $^{15}N–NH_4^+$ by different pools, 50 hours after labelling two grasslands: a productive, fertilized grassland (improved) and an unproductive, unfertilized grassland (unimproved.) Data are expressed as % added. (Data from Bardgett et al. 2003.)

further studies to determine the mechanisms that regulate the capacity of microbes to sequester N.

Not all ecosystems accumulate N when subject to N deposition; rather, some reach N saturation very rapidly resulting in significant amounts of N export in drainage waters. This has been shown to be the case in hardwood forests of northern United States, for example. Here, 8 years of experimental NO_3^- additions (at 2.5 times ambient N deposition) to a number of sites dramatically increased leaching losses of NO_3^- and DON, indicating rapid N saturation of this ecosystem (Pregitzer et al. 2004). This is in contrast to the fore-mentioned studies of UK semi-natural grasslands where significant amounts of added N were retained in soil organic matter, thereby reducing N losses (Phoenix et al. 2004). It is not entirely clear why some ecosystems are more susceptible to N saturation than others, but one suggestion is that it depends on initial levels of N availability (Aber et al. 1998). For example, Gundersen et al. (1998) showed that forests with the highest risk of N saturation and NO_3^- leaching had surface organic horizons with C : N ratios of <25, whereas a survey of 181 forests across Europe by MacDonald et al. (2002) showed that forests with organic horizons of C : N <25 and soil pH 4.6, and atmospheric N inputs >3 g N m^{-2} yr^{-1}, were most susceptible to N saturation. Similarly, it is well established that fertile grasslands that receive regular additions of fertilizer N are especially prone to N saturation and N export, which contrasts with low fertility grasslands where microbes sequester a large proportion of added N, which is then transferred to live plant and stable organic matter pools (Bardgett et al. 2003) (Fig. 6.12). In other words, inherently N-rich soils receiving high atmospheric N deposition are very susceptible to N saturation and large losses of N via leaching (Pregitzer et al. 2004).

The ability of soils to retain N is likely to be strongly affected by changes in plant species composition that accompany increased N deposition, which potentially lead to significant feedbacks to the decomposer community and soil N cycling. A good example of how N-driven shifts in plant community structure influence N cycling comes from moist meadow communities of alpine tundra of the Rocky Mountains, USA. These communities are especially susceptible to N deposition, since they receive relatively large inputs of N from melting snow, an important reservoir of wintertime N deposition (Bowman 1992). Furthermore, these plant communities are cohabited by both the N-limited grass *Deschampsia caespitosa* and the N-insensitive forb *Acomastylis rossii*, which exert markedly different influences on soil N cycling: typically, rates of N mineralization in soils under *Deschampsia* are 10-fold greater than beneath *Acomastylis* (Steltzer and Bowman 1998). Therefore, the replacement of *Acomastylis* as a dominant in moist meadow by *Deschampsia* under elevated N deposition, leads to a strong positive feedback, increasing rates of N mineralization and leaching of N from soil, leading to greater N export from alpine catchments (Welker et al. 2001).

This kind of positive feedback is probably mirrored in many situations where N deposition promotes the dominance of fast-growing plant species that produce N-rich plant litter, which is easily decomposed by microbes in soil, leading to rapid N mineralization and potential for N loss (Fig. 6.13). Furthermore, the potential for N export is likely to be further promoted by associated shifts in soil food-web structure towards bacterial dominance. As discussed in Chapter 4, the encouragement of bacterial-dominated food webs by nutrient enrichment promotes fast and 'leaky' nutrient cycles with a high potential for nutrient export to adjacent ecosystems. As a result of these feedbacks to increasing N, it has been proposed that responses to N deposition will be non-linear (Fig. 6.14) since the export of N from soils will be accelerated as a result of species replacement, owing to enhanced N cycling in soil (Welker et al. 2001).

Before leaving the subject of N deposition, it is important to highlight one other area of research that is receiving increasing attention, that is, the question of how atmospheric N deposition influences the production of DON and its loss from polluted soils. Until recently, fluxes of DON were largely ignored, with emphasis being placed on the cycling and loss of inorganic N. However, a number of studies reveal that several ecosystems facing N deposition, including grassland and forest, are subject to high amounts of DON loss in leachates relative to inorganic N (McDowell et al. 1998; Phoenix et al. 2004; Pregitzer et al. 2004). A good example of this is in hardwood forests of North America, where experimental N additions increased DON export six-fold (Pregitzer et al. 2004). Similarly, Phoenix et al. (2004) were surprised by the large amount of DON exported in leachate from semi-natural grasslands subject to N deposition. This

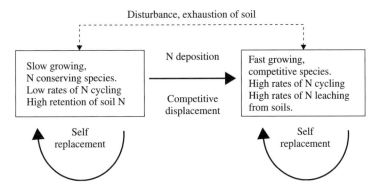

Fig. 6.13   Conceptual model showing the generalized pattern of plant species self-replacement in communities with low and high N availability. The model shows that in the face of greater N deposition, N conserving species of infertile environments will be replaced by more competitive species, with traits that promote more rapid N cycling in soil. Under such conditions, the new state would be preserved, even in the absence of additional N inputs, until N reserves in soil are exhausted when species adapted to low N conditions will re-establish. (Redrawn with permission from Oxford University Press; Welker et al. 2001.)

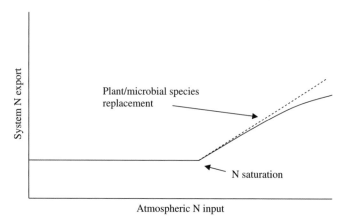

Fig. 6.14   Hypothetical, non-linear response of system N exports to increasing N deposition with (dashed line) and without (solid line) plant and microbial species replacements. As N deposition increases, a point is reached where system N export begins to increase during the growing season (N saturation.). As N availability continues to increase, species replacements result in greater N export owing to greater rates of N cycling and lower N retention (Redrawn with permission from Oxford University Press; Welker et al. 2001.)

high loss of DON from polluted soils has been attributed to effects of N deposition on the production of organic substrates and their processing by soil microbes, and to incomplete degradation of lignin as a result of inhibition of lignolytic enzymes, leading to increased soil concentrations of phenolics and greater production of DON in polluted soils (Waldrop et al. 2004). Further studies are clearly needed in this area to determine the mechanisms controlling the production and loss of DON from polluted soils and the significance of this form of N loss for adjacent ecosystems.

6.4   Invasive species

Over recent decades, the invasion of exotic species into habitats has emerged as an important cause for species decline and native habitat degradation (Drake and Mooney 1989; Vitousek et al. 1997; D'Antonio et al. 2001). The scale of this problem is enormous: invasive species are almost everywhere on Earth and some believe that they will become the leading cause of habitat degradation (D'Antonio et al. 2001). On many islands, for example, more than half the plant species are non-native, and in many continental areas the figure is 20% or more (Vitousek et al. 1997). In these situations, the invasion of exotic species into natural ecosystems can negatively affect native species directly, through competition, predation, or disease, or indirectly by altering ecosystem processes such that native species begin to die out. This section considers how both above-ground and below-ground invasive organisms can impact on indigenous soil biota and soil processes, and how this results in strong feedback to nutrient dynamics and ecological processes that influence the ability of ecosystems to sustain populations of native species.

6.4.1 Above-ground invasive organisms

Above-ground invasive species that have the greatest impact on soil biological and ecosystem processes are likely to be those that have traits that are qualitatively most different from the native species in invaded areas (Chapin et al. 1996*b*). There are several routes through which above-ground invasive species impact on ecosystem processes, but those most likely to cause the greatest effects are those that: (1) introduce a new trophic level to the ecosystem; (2) alter the rate of resource supply to the system; or (3) alter the disturbance regime (Vitousek 1990). This framework, along with specific examples, will be used below to show how invasions impact on below-ground properties in an ecosystem context.

An especially good example of the ecological impacts of the invasion of a whole functional group of alien organisms into an ecosystem concerns the introduction of large herbivorous mammals into New Zealand, where they were previously absent. Several species of large browsing mammals were introduced to New Zealand between the 1770s and 1920s, the most ecologically significant being European red deer and feral goats. As discussed in Chapter 5, these browsing mammals have had consistent adverse effects on the productivity and diversity of understorey vegetation in forests, promoting the growth of unpalatable species with poor litter quality, such as ferns, shrubs, and small-leaved monocotyledonous species, at the expense of fast growing palatable broad-leaved shrubs with higher litter quality (Wardle et al. 2001). Despite the consistency of these trends, the response of below-ground biota has been found to be far less predictable: most groups of smaller-bodied soil organisms, such as microfauna and microflora, responded both positively and negatively to browsing, depending upon the site considered. Similarly, ecosystem processes driven by soil biota, such as soil C mineralization and C and N sequestration, showed these types of context-dependent responses. What this example illustrates is the important, albeit idiosyncratic, effects that introduced browsing mammals can have on the decomposer subsystem, which in the long term may affect rates of nutrient supply from the soil and ultimately the nutrition, productivity, and composition of the forest ecosystem. It is also likely that these sorts of ecosystem changes are irreversible in the long term (Coomes et al 2003). Although the issue of how alien browsing mammals affect below-ground properties has been seldom investigated, a number of studies worldwide have documented important effects of introduced mammals on vegetation composition (see review by Vazquez 2002). It is therefore likely that there are many situations in which they may exert significant impacts on decomposer communities and their activities, and hence the long-term performance of the ecosystem (Wardle and Bardgett 2004*a*).

Another route by which invasive species influence ecosystem properties that sustain native species is through modifying the availability of growth-limiting resources. A classic example of this comes from the work of Vitousek et al. (1987) and Vitousek and Walker (1989) who studied the effects of the invasion of the N-fixing actinorhizal shrub *Myrica faya* into young, N-limited *Metrosideros polymorpha* forests in Hawaii, which previously lacked symbiotic N fixers. Immigrants introduced this species into the Hawaiian Islands from the Azores and Madeira in the 1800s as an ornamental or medical plant. Owing to its prolific seed production and effective dispersal by exotic birds, and rapid growth rate in open-canopied sites, it has now spread into young, N-limited natural forests of Hawaii. This has led to increased soil N inputs by some four-fold as a result of its N-fixing activities, hence greatly increasing the size of the N pool and soil N availability (Table 6.3). This enrichment of native forest soils with N, the one nutrient that they lack, has potentially dramatic consequences for these ecosystems. Not only does it increase their productivity, but also it can potentially feedback by creating a more favourable environment for invasion by a broader range of exotic species (Vitousek and Walker 1989). Indeed, sites with fertile soils tend to be invaded more successfully by a broader range of plant species than are sites with infertile soils on Hawaii (Gerrish and Mueller-Dombois 1980) and elsewhere (Pickard 1984). Overall, this example demonstrates how biological invasions can greatly alter soil properties in such a way that substantially changes soil nutrient availability and ecosystem productivity, and also alters the future vegetation composition of natural ecosystems.

Table 6.3 Data showing the effects of invasion by the N-fixer *Myrica faya* on N inputs into *Metrosideros polymorpha* forests ('ohi'a lehua), Hawaii.

Source	Invaded	Non-invaded
Fixation by *Myrica faya*	18.5	0.2
Native N fixation		
Lichens	0.02	0.06
Litter	0.12	0.16
Decaying wood	0.05	0.03
Precipitation		
Inorganic N	1.0	1.0
Organic N	2.8	2.8
Total inputs	22.5	4.2

The table shows annual N inputs (kg ha^{-1}) into forest sites that are densely invaded (>1000 ha^{-1}) and where *Myrica* has not yet invaded to any great extent.

Source: Adapted from Vitousek and Walker 1989.

Invasive organisms can also influence ecosystems and soil properties by modifying disturbance regimes. There are numerous examples of where introduction of fire-enhancing invasive grasses to ecosystems has greatly increased the frequency of disturbance by fire (D'Antonio and Vitousek 1992; Brooks et al. 2004). For example, several alien grasses have been implicated in increasing fire frequency and/or intensity in Hawaii, the best documented case being the invasion of the C_4 perennial grass *Schizachyrium condensatum* into areas of submontane woodland. This grass grows into the canopies of native shrubs, creating a continuous layer of fuel, greatly increasing the occurrence and intensity of fire. This generally has the effect of killing native trees and shrubs, enabling *S. condensatum* and other grasses to invade burned areas, further increasing the risk of fire. In other words, grass invasion sets in motion a feedback cycle that leads from a non-flammable, mostly native-dominated woodland to a highly flammable, low diversity, alien-dominated grassland (D'Antonio and Vitousek 1992). Another example of a widespread invader that has caused tremendous changes in fire regimes and other ecosystem properties is the alien annual grass *Bromus tectorum* in western North America. Its invasion across this vast landscape has increased fire frequency to the point that native shrub-steppe species cannot recover (Brooks et al. 2004). The invasion of *Bromus rubens*, another non-native to North America, in to the Mojave Desert poses a similar threat to fire regimes and native plants (Brooks and Esque 2002). Although not tested, such changes in plant community structure and fire frequency will undoubtedly have dramatic effects on soil organisms and their activities, which will feedback to the plant community in terms of altered nutrient availability and process rates (Fig. 6.15).

6.4.2 Below-ground invasive organisms

Very little study has been done on invasions of the soil community and its effect on ecosystem processes related to nutrient cycling. Work that has been done has focussed on invasions of large fauna, especially of earthworms that are known to profoundly influence soil biological and physical properties and ecosystem function (Chapter 3). Many of these studies have been done in northern temperate forests of North America (e.g. Alban and Berry 1994; Scheu and Parkinson 1994) where native earthworms were eliminated during the last glacial period and were slow to recolonize after the glaciers receded (James 1995). Consequently, most earthworms that occur in soils of these ecosystems are invasive species from Europe and Asia, which are now colonizing temperate forest sites across a large geographic area of North America. The earthworm species involved have been called the 'peregrine' species, which represent a small portion of the overall earthworm diversity, but are characterized by their ability to colonize new habitats and spread rapidly, and tolerate a wide range of environmental conditions (James and Hendrix 2004).

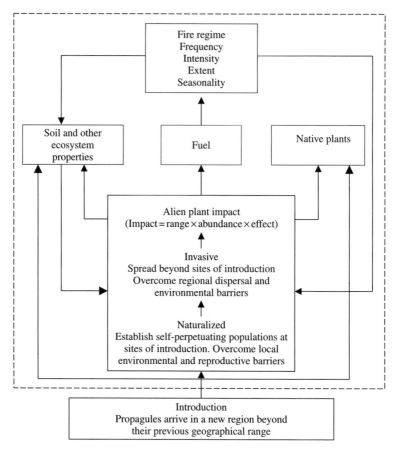

Fig. 6.15   Schematic diagram of the invasive plant-fire regime cycle (Brooks et al. 2004.)

Bohlen et al. (2004*a*) conducted a large-scale research project to address the hypothesis that earthworm invasions of these north temperate forests have large consequences for nutrient retention and uptake in these ecosystems (Fig. 6.16). They found that the most dramatic effect of earthworm invasion was the loss of forest floor, which exposed roots and greatly altered the location and nature of nutrient cycling (Fig. 6.17) (Bohlen et al. 2004*b*). At one site, earthworm invasion led to mixing of the forest floor into the mineral soil, leading to a net loss of approximately 25% of soil profile C to a depth of 12 cm, mostly due to enhanced mineralization of forest floor C (Bohlen et al. 2004*b*). Despite this loss of C, there was no loss of N from the soil profile, or any effect of invasion on microbial processes of N cycling, such as N mineralization or nitrification. However, large amounts of N were conserved in the mineral soil following invasion, presumably due to microbial immobilization of N in these deeper soil horizons (Groffman et al. 2004). This view was supported by the finding that while earthworm

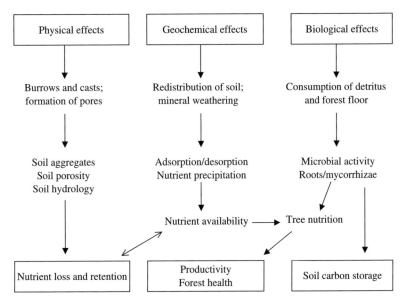

Fig. 6.16 Schematic diagram illustrating the main mechanisms by which invasive earthworms influence forest biogeochemistry. (Redrawn with permission from Ecosystems Vol. 7, 2004, Springer; Bohlen et al. 2004*a*.)

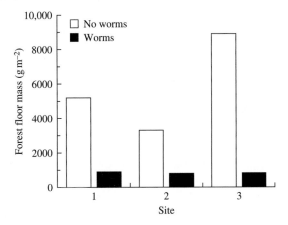

Fig. 6.17 Forest floor mass in paired earthworm-invaded and reference plots at three locations at Cornell University's Arnot forest, western New York State, North America. The forest was logged in the late nineteenth century and has developed with minimal disturbance since then. It is dominated by six species characteristic of northern hardwood forest that is, sugar maple, red maple, beech, white ash, basswood, and hemlock. Data show significant reduction in forest floor mass as a result of invasion by exotic earthworms. (Data from Bohlen et al. 2004*b*.)

invasion led to a greatly reduced microbial biomass in the depleted forest floor, it increased significantly microbial growth in the deeper mineral soil, which could have acted as a sink for N, conserving N in the soil profile (Groffman et al. 2004). There was evidence of increased nutrient uptake efficiency by fine roots in the form of a reduction in root respiration and fine root biomass in invaded forest stands (Fisk et al. 2004). While not directly measured, it was hypothesized that enhanced nutrient uptake by roots was a result of the positive effects of earthworm mixing on mycorrhizal and root turnover, and rhizodeposition, resulting in increased C supply to heterotrophic microbes (Fisk et al. 2004). Overall, these results indicate that exotic earthworm invasion—which is predicted to accelerate under climate change—greatly influences the structure and function of northern forests, and could increase the potential for C loss from these ecosystems (Bohlen et al. 2004a). Further information on the potential effects of invasive earthworms on temperate forests can be found in an excellent review paper by Bohlen et al. (2004c).

Invasive earthworms can also impact on ecosystems by competing with native populations and by altering soil physical properties that affect material movement through soil, such as aggregation and pore formation. A good example of this comes from the work of Chauvel et al. (1999) who showed that the conversion of Amazonian forest to pasture reduced the original soil macrofaunal community by 68% (from more than 160 species to less than 40 species) and was coupled with the invasion of the soil by the earthworm *Pontoscolex corethrurus*, an aggressive exotic colonist. This invasive earthworm, which is now dominant in many farming systems throughout the tropics (Lapied and Lavelle 2003), produces compact casts that accumulate in the surface soil, forming continuous crusts that prevent water infiltration and plant growth. Consequently, it is causing a great loss in the earthworm biodiversity in the tropics (Fragoso et al. 1999a) with associated problems owing to the deterioration of soil structure.

Native earthworm populations and their beneficial effects on soil properties can also be negatively affected by invasions of predatory macrofauna into new territories. The accidental introduction of the New Zealand flatworm *Arthurdendyus triangulata* into Northern Ireland and the Faeroe Islands in the 1960s, and its subsequent spread throughout Britain (Boag et al. 1995, 1997) is a good example of this. Observational evidence suggests that as agricultural grasslands become invaded by flatworms, earthworm populations decline substantially, sometimes below detectable limits (Blackshaw 1990). It has also been suggested that flatworm-induced reductions in earthworm densities result in impeded soil drainage, alterations in plant community, and losses of soil vertebrates such as moles (Boag 2000; Boag and Yeates 2001). Although not tested experimentally, such changes are likely to have significant consequences for other decomposer organisms and processes of decomposition and nutrient supply, resulting in strong feedback to nutrient dynamics and ecosystem processes. Another example of how an invasive predator of soil biota indirectly alters ecosystem properties comes from subantarctic Marion Island. As discussed in Chapter 3, the soil-borne larvae of a flightless moth *Pringleophaga marioni* process some 1.5 kg plant litter or peat $m^{-2} yr^{-1}$, thereby stimulating N and P mineralization 10-fold and 3-fold, respectively (Smith and Steenkamp 1990). However, the introduction onto the island of the house mouse *Mus musculus*, which feeds on the moth larvae, has reduced annual litter turnover by some 60% of that originally processed by the moth (Smith and Steenkamp 1990). According to Smith and Steenkamp (1990), if mouse numbers continue to increase, rates of peat accumulation will increase which will, in turn, dramatically alter the hydrological regime and vegetation of the island.

6.5 Land use transformation

The use of land for agriculture, commercial forestry, and other services represents the most substantial human alteration of the Earth system, and at present represents the primary driving force in the loss of biological diversity and habitat degradation worldwide (Vitousek et al. 1997; Wilson 2002). For example, arable agriculture and urban development occupy 10–15% of Earth's land surface, and a further 6–8% has been converted to intensively managed pasture. The majority of Earth's remaining vegetation is either grazed by domestic animals or harvested for wood products to varying degrees (Vitousek et al. 1997). Furthermore, it has been estimated that some 17% of Earth's vegetated land has been degraded, largely through erosion owing to deforestation and farming practices that lead to over-grazing (Kaiser 2004). All these human activities, and especially the conversion of natural systems to farmland and forestry, have significant consequences for soil biota, generally decreasing their abundance and diversity (Chapter 2). Also, changes in the intensity of land use, for example, for the production of agricultural crops and livestock, have significant consequences for soil biota that ultimately influence the functioning and sustainability of these systems. This section is concerned with how variations in management intensity of agricultural land affect soil biota and some of the processes that they regulate.

Studying the effects of changes in agricultural management intensity on soil biota and the processes that they regulate has for many years been a major theme of soil ecology. This is largely because the detection of responses of soil biota and their activities to changes in farming practice offers a means of identifying either deleterious or beneficial effects of management on the functioning of soil (Brussaard et al. 1997). There are many different management components involved in the intensification of agricultural management, including increased use of fertilizers and pesticides, increased disturbance of soil from tillage, vehicle traffic and grazing pressure, and reduced diversity of resource inputs to soil from the use of single crop species. Since it is difficult to separate the individual effects of these components on soil biological properties, most studies in this area have used a 'systems' approach whereby they simply compare agricultural systems of different levels of intensification (Bardgett and Cook 1998).

A general pattern that emerges across studies of farming systems is that intensification alters the relative importance of bacterial and fungal-based pathways of decomposition (*sensu* Coleman et al. 1983). What appears to occur is that intensive management, characterized by high levels of disturbance and fertilizer use favour bacterial-based energy channels in which microfauna play an important role. In contrast, low input systems, with less disturbance and greater inputs of more complex organic matter, favour fungal-based energy channels and greater involvement of earthworms and microarthropods. This has been shown to be the case in many comparative studies of intensive versus low input farming systems, a classic example being the comparisons of conventional tillage and non-tillage systems at Horseshoe Bent, Georgia, USA. This study showed that long-term conventional tillage, which greatly reduces soil organic matter content, has a detrimental effect on larger organisms, especially earthworms, but favours bacterial-feeding nematodes relative to fungal-feeding fauna (Hendrix et al. 1986). This shift in soil food-web structure is also reflected in the contribution of different organism groups to total soil C loss. Beare (1997) calculated that bacteria, protozoa, and nematodes made greater individual contributions to total respiratory loss of C from the conventional tillage than in the non-tillage systems, especially in the warm season when differences in bacterial biomass between systems were greatest (Table 6.4). Using biocide treatments to eliminate different groups

Table 6.4 Differences in the calculated respiratory losses of C from microbial and faunal groups in a subtropical Ultisol under conventional tillage (CT) and no-tillage (NT) management

Soil biota	Annual total	
	NT	CT
Losses of C (% of total)		
Fungi	22.3	19.3
Bacteria	62.8	72.8
Protozoa	2.5	3.7
Nematodes	0.16	0.25
Microathropods	0.11	0.03
Encytraeids	0.52	0.40
Earthworms	11.2	3.5
Total losses of C (kg ha^{-1} season^{-1})		
Total respiration	7486	8334
Decomposition	6980	8250

Source: Adapted from Beare 1997.

of soil organisms, Beare et al. (1997*a*) also showed that fungi contributed more to decomposition of crop residues in the non-tillage than the tillage system, and that fungal population dynamics and N immobilization were strongly regulated by fungal-feeding microarthropods in the non-tillage system. Mycelial fungi also contributed significantly to the structural stability of non-tillage soils by forming stable macro-aggregates (Beare et al. 1997*a*).

Studies of gradients of management intensity in temperate grasslands also reveal that soil foodwebs of intensive systems, characterized by high levels of fertilizer use, high grazing pressures, and reduced soil organic matter content, consistently have bacterial-dominated soil food webs, whereas soils of traditionally managed, unfertilized grasslands have fungal-dominated food webs. For example, Bardgett and McAlister (1999) analysed soils from a wide range of meadow grassland sites in northern England, along a gradient of long-term management intensity, and showed that fungal biomass and the fungal-to-bacterial biomass ratio (measured using PLFA) were consistently and significantly greater in traditionally managed, unfertilized, than intensively managed, fertilized grasslands (Fig. 6.18*a*). Similarly, Grayston et al. (2001*a*, 2004) examined soil microbial community structure across gradients of grassland management intensity at a number of sites within different biogeographic regions of the British Isles; they found that fungal biomass and fungal-to-bacterial biomass ratios were consistently greater at the low intensity end of the management gradient (Fig. 6.18*b*). These findings also relate well to those of Yeates et al. (1997*a*) who compared a series of intensively managed grasslands with adjacent sites that had been managed under organic prescriptions (i.e. no inorganic fertilizers) for several years in mid-Wales, UK. These authors found that the nematode community was larger and more species-rich in the organically managed grassland systems and fungal-feeders were twice as abundant as in conventionally managed soils. This change was also mirrored in a shift in the soil microbial community towards fungal dominance (Yeates et al. 1997*a*). Similar impacts of management on the character of soil food webs have been detected in studies of Dutch grassland where the density and species diversity of soil microarthropods is much lower in fertilized, high-input grasslands than in unfertilized, low-input sites (Siepel 1996). Furthermore, fungal-feeding grazers dominate the microarthropod community in the low-input sites, whereas in the high-input sites these organisms may be replaced by opportunistic bacterial-feeders (Siepel 1996).

It is not entirely clear why bacterial-dominated food webs develop so rapidly under intensive management, but it has been proposed that it is the result of the combined effects of a number of factors, including: changes in plant species composition and the dominance of fast growing species that produce litters and rhizodeposits that encourage bacterial

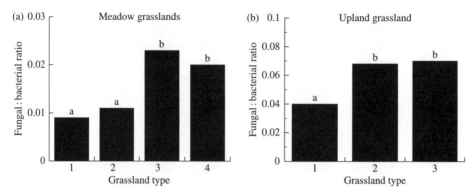

Fig. 6.18   Effects of management intensity on fungal-to-bacterial biomass ratios measured using PLFA. (a) Gradient of declining management intensity on meadow grasslands in northern England, characterized by long-term reductions in fertilizer use and livestock density, classified as: (1) Improved meadow; (2) Very modified meadow; (3) Slightly modified meadow, and; (4) Unmodified meadow. (b) Gradient of three upland grassland types of varying management intensity across 10 sites in different biogeographic areas of the United Kingdom, from: (1) improved *Lolium-Cynosurus* grassland; (2) semi-improved *Festuca-Agrostis-Galium* grassland, *Holcus-Trifolium* subcommunity, and; (3) unimproved *Festuca-Agrostis-Galium* grassland. Values are means and bars with the same letter are not significantly different. (Data from Bardgett and McAlister 1999 and Grayston et al. 2004.)

growth (Chapter 4); reductions in the complexity and heterogeneity of organic matter inputs; reductions in habitat complexity of the soil environment; and increases in physical disturbance of soil from vehicle traffic, grazing, and tillage. Reductions in fungal growth resulting from intensification have also been attributed to direct inhibitory effects of fertilizers on the growth of decomposer fungi (Garbaye and Le Tacon 1982; Donnison et al. 2000*b*) and AM fungi (Sparling and Tinker 1978; Johnson et al. 2003*c*). Agricultural practices associated with intensification, such as ploughing and the addition of phosphate fertilizer, are also known to greatly suppress mycorrhizal infection in crop plants, through disruption of mycelial systems in the case of the former, and reduced mycorrhizal-dependency in the case of the latter (Leake et al. 2005).

While intensification of agricultural management can induce rapid replacement of fungal-dominanted foods webs by bacterial-dominated ones, with accompanied loss of species, the reversal of this shift in community composition is a long-term process. For example, in Siepel's (1996) study, abandoned high-input sites still lacked fungal-feeding mites after 20 years of management for nature conservation as a result of the low population growth and dispersal rate of these species (Siepel 1996). Similarly, Scheu and Shultz (1996) found a very slow change in the oribatid mite community following cessation of cultivation, despite dramatic changes in the plant community and rapid recolonization of uncultivated soils by the more mobile macrofauna. The same appears to be the case for microbial communities. For example, Smith et al. (2003) found that it took nine years of re-instatement of traditional, low input management to intensively managed grassland to detect any recovery in the fungal-to-bacterial biomass ratio of the microbial community, which had previously been depressed by decades of intensive management. Earthworms, by virtue of their limited dispersal ability, also slowly recolonize favourable habitats that result from land-use change as illustrated by the absence of earthworms in the Dutch polders reclaimed between 1942 and 1975 (Brussaard et al. 1996).

To summarize the response of soil biota to agricultural land management, two patterns seem to emerge. First, natural ecosystems and low-intensity farming systems appear to have more diverse, fungal-dominated decomposer food webs with larger organisms than do intensive agricultural systems. Second, changes in soil food webs are not necessarily reversible in the short term. The net effect of these changes in soil food webs for the functioning of agricultural soils is not entirely clear. However, it is probable that an increased role of bacterial-dominated energy channels with increasing intensification will be associated with faster, more leaky nutrient cycling, with more losses of nutrients and C in water and gases to the surrounding environment. Conversely, low intensity management systems encourage more diverse, fungal-dominated food webs that are more similar to those of natural ecosystems. Owing to more efficient nutrient cycling that is more in line with plant nutrient demand, such systems will potentially require fewer management inputs due to greater reliance on ecosystem self-regulation (Altieri 1991; Yeates et al. 1997*a*; Bardgett and McAlister 1999).

6.6   Conclusions

All the human-induced global changes discussed in the chapter are ongoing, and in many cases they are accelerating. Furthermore, they are operating together, meaning that Earth's soils are often subject to a suite of global change phenomena that have the potential to substantially alter their structure and function. It is clear from the evidence presented in this chapter that soils and their biota are substantially modified by global changes. Also, it is evident that below-ground responses to global change ultimately feedback to nutrient dynamics and the plant community, thereby indirectly influencing the effect of global changes on ecosystem structure and function. In other words, effects of global change on ecosystems consistently involve linkages between above-ground and below-ground communities that are often specific to the ecosystem where they operate (Wardle et al. 2004*a*). It is also apparent that the ability of ecosystems to withstand, or tolerate global change phenomena in the long term is related to soil biological properties, especially those properties that interact directly or indirectly with plants. This is clearly demonstrated by the common observation that net effects of global changes on ecosystems depend on which groups of soil biota are affected, and also the extent to which microbes in soil are nutrient-limited. In view of this, future studies need to take account of such relationships between plant and soil communities for us to understand fully the response of ecosystems to a changing world.

In this chapter, global change phenomena have been considered individually. However, in nature they operate simultaneously, leading to strong interactive effects of global changes on soil biological and ecosystem properties. Remarkably little is known about the nature and importance of interactive effects of global change phenomena, but they are likely to be highly significant in terms of the net effect of humanity on soils and ecosystems. For example, the ability of soils to sequester C in a high CO_2 world is likely to be altered substantially by N deposition and the activities of invasive species, and also by factors such as land-use change. Understanding how linkages between below-ground and above-ground communities are influenced by simultaneous global changes is thus a key priority for the future.

7 Conclusions

The study of soil biodiversity and its consequences for ecosystem properties is a relatively new field of ecology. While it has been known throughout history that soil organisms are integral to soil fertility, it is only in the last few decades that ecologists have started to explore the vast diversity of biota in soil and their functional significance for plant communities and ecosystem processes. This is, in part, due to technological advances that have enabled scientists to extract and characterize microbial diversity of soils and assess its function. However, it is also due to an increasing recognition by ecologists, who have traditionally focused on organisms that live above-ground, of the importance of below-ground diversity for the function of terrestrial ecosystems.

One striking characteristic feature of soil biological communities is that they are incredibly complex and diverse, and thus represent a major reservoir of global biodiversity. The extent of the diversity of microbial communities has only recently become apparent as a result of the use of molecular techniques, which reveal an astonishing diversity of bacteria and fungi in soil, much greater than previously thought possible (Curtis et al. 2002). Indeed, it is probably fair to say that soil contains more species than any other environment on Earth, and it is the most complicated, physically heterogeneous habitat, with a bewildering physical complexity that occurs across a range of spatial and temporal scales. As a result there is extreme spatial and temporal variability in community-level biological properties of soils (Ritz et al. 2004). This high biological and physical diversity, however, are inextricably linked; high biological diversity results from high heterogeneity since it provides unrivalled potential for resource and habitat specialization of soil organisms, leading to avoidance of competition and hence a high level of species coexistence. Scientists are now developing new techniques to study these complex soil communities in their natural environment. For example, new visualization and geostatistical tools are being developed for interrogating the inner architecture of soil, enabling soil organisms to be observed *in situ* in their natural habitat (Nunan et al. 2001; Young and Crawford 2004). These studies are greatly advancing our understanding of the distribution of microbial communities in soil, revealing that they are highly ordered, with individual communities being aggregated in specific ways in response to spatial heterogeneity imposed by the habitat (Ettema and Wardle 2002; Young and Crawford 2004).

Information on the patterning of soil biodiversity is relatively scarce, at least compared to what is know about above-ground communities. However, certain patterns are emerging and it is clear that, as with plant communities, patterning of soil biodiversity occurs across a range of temporal and spatial scales. Large-scale patterns of soil biodiversity that occur across regional scales relate primarily to dispersal and to higher order climatic factors, which determine rates of primary productivity and resource availability, and also the activity of decomposer organisms. An important question here concerns the global distribution of soil organisms, and the idea that small microbes, whose dispersal is not limited, are everywhere (Finlay 2002). This is a contentious issue, but it is clear that large-scale dispersal mechanisms play a central ecological role in structuring soil communities and that information relating to this is lacking. Patterning at smaller spatial scales, for example, within grassland or forests, or agricultural fields, is also highly complex and appears to relate mostly to patterns of resource availability and heterogeneity created by litter, animal wastes, roots and other organic materials, and patterns of root carbon flow. Variations in soil structure that affect the physical dimensions of habitable space also contribute greatly to the heterogeneity of soil. Finally, as discussed above, patterning of diversity at the microscale appears to be related to heterogeneity at the level of the soil particle, where steep environmental gradients in moisture and oxygen availability can act as important determinants of microbial diversity. In sum, variations in spatial and temporal heterogeneity of the soil environment, from the scale of the soil particle to the soil horizon, coupled with the extreme specialization of the soil biota, act as the major determinant of patterns of soil biodiversity.

On the subject of patterns of soil biodiversity, an important conclusion that does emerge from this book is that soil biological communities are very susceptible to disturbance, especially those caused by human intervention, which very often lead to significant loss of soil species. This book includes many examples of where the conversion of natural lands to agriculture has led to dramatic declines in the diversity of both fauna and microbes. This is perhaps most obvious for large fauna, such as termites, which show an almost linear decline in species richness with increasing agricultural intensification in the tropics (Eggleton et al. 2002). However, the diversity of smaller organisms, such as nematodes and microbes, is also often impaired by conversion of natural lands to agriculture, or by increases in the intensification of agriculture. It is also evident that the loss of species from soils as a result of disturbances is not a random process, in that certain species are more susceptible than others to disturbance (Siepel 1996). Also, of particular concern is that species that are lost from soil due to agricultural disturbance or intensification often are very slow to recover when management is reversed. This was shown by Siepel (1996) who found that intensive farming of temperate grasslands led to a rapid loss of fungal-feeding mites from soil, but that these did not recover even after 20 years of management for nature conservation. This clearly has implications for management of soils and their diversity, implying that changes in soil biological communities resulting from agricultural or other human-induced disturbances are not always reversible within medium timescales.

Moving on to the roles of soil biological communities, a dominant theme of this book is the importance of biotic interactions, including trophic interactions between organisms that live in soil and those that occur between soil biota and the plant community. There is now very strong evidence that these biotic interactions are fundamental to understanding the functional role of soil biota at the ecosystem level. While microbial processes of nutrient mineralization are of central importance for plant nutrient supply, there is much evidence to indicate that trophic interactions between microbes and their animal predators strongly influence rates of nutrient supply to plants, thereby influencing plant productivity. There is also evidence that trophic cascades, involving top predators that feed on microbial-feeding fauna, can be of functional significance for shaping microbial communities and influencing plant nutrient supply. Also, evidence is accumulating of important non-nutritional effects of animal–microbial interactions on plant growth, in that animals indirectly influence root growth by influencing bacterial production of plant growth-promoting hormones. Other types of biotic interactions that influence ecosystem form and function include mutualistic associations with plants, such as mycorrhizal fungi and N fixers, and the interactions between soil-dwelling plant pathogens, such as fungal pathogens and herbivores, and plants. All these biotic interactions, together with impacts of soil organisms on the biophysical nature of decomposing material and of the soil environment, collectively act as the drivers of soil biological functions that drive ecosystem processes of decomposition, nutrient supply, and plant productivity.

The question of whether changes in soil biodiversity have consequences for ecosystem function is another major theme of this book. From the studies that have been done to date, the message that emerges is that changes in the abundance of particular species and the nature of trophic interactions are more important drivers of ecosystem function than is diversity *per se* (i.e. the number of species). This is consistent with studies of above-ground communities which point to the role of vegetation composition and dominant plant species as a major driving force of ecosystem function at local scales (Grime1997; Hooper and Vitousek 1997; Wardle et al. 1997*a*). It is also consistent with the view that there is a high degree of redundancy within soil communities, as evidenced by the prevalence of omnivory in soil food webs (Scheu and Faka 2000; Maraun et al. 2003), and the knowledge that some species are more redundant than others (Laakso and Setälä 1999*b*). It is important to note, however, that nearly all the studies that these views are based on are on soil animals, which lend themselves to manipulation under experimental conditions more than do soil microbes. Much less is known about the role of microbial diversity as a driver of soil functions, although a general view is that redundancy within microbial communities varies for different functional groups of microbes (Schimel 1995). In other words, influences of microbial community composition and diversity are most likely for processes that are physiologically or phylogentically 'narrow', such as N-fixation or nitrification, whereas 'broad' processes such as N mineralization are insensitive to microbial community composition. This view, however, has been challenged, with scientists now arguing that 'broad' processes of mineralization and immobilization can be 'aggregated' into individual components, based on specific enzyme activities and spatial distribution of microbes in microsites, that are sensitive to microbial community composition (Schimel et al. 2005). Again, this emphasizes the important need for better understanding of the spatial organization of microbial communities within soil and the need to simultaneously disentangle the specific functions of microbes involved in important soil functions. Recent methodological developments in the use of stable isotopes and their assimilation into biomarkers are now enabling us to tackle such questions (e.g. Radajewski et al. 2000).

An important message that is emerging is that relationships between soil biotic interactions and ecosystem processes are strongly context dependent, in that their nature and significance varies in space and time. In other words, soil biodiversity will be an important driver of ecosystem processes in some situations, but not in others where abiotic controls will dominate. Not much is known about the relative importance of abiotic and biotic controls on soil processes, but a handful of studies show that in certain ecosystems abiotic factors such as moisture and temperature act as more important determinants of soil biological processes than do changes in the structure and biomass of soil communities (Gonzalez and Seastedt 2001; Liiri et al. 2002). The strong implication from these studies is that while diversity and community composition of soil organisms might be important regulators of ecosystem processes at the local scale, abiotic factors can be stronger regulators at larger, regional scales. Overall, much is still to be learnt about the functional role of soil food web interactions, especially of microbes, and also the relative importance of these biotic and abiotic forces as drivers of ecosystem function. This is a major research challenge for the future.

This book illustrates that plant and soil communities are mutually dependent on one another, and that feedback between plant and soil communities act as an important driver of ecosystem structure and function. As discussed in Chapter 4, plants provide C and other nutrients to the decomposer community, but plant roots also act as a host for many soil organisms, such as herbivores, pathogens, and symbionts. The soil biota, in turn, influence plant communities indirectly by recycling dead plant material and making nutrients available for plant use, and directly through the action of the root-associated organisms which selectively influence the growth of plant species, thereby affecting plant productivity and community structure. Once again, the nature and significance of these interactions between plants and soil biota appear to be highly context dependent and involve a wide range of multi-trophic interactions. However, conceptual models are being developed in this area, which offer exciting insights into the functional importance of feedback mechanisms as drivers of community change (Wardle et al. 2004a). It is becoming evident, for example, that certain plant species have the capacity to encourage the accumulation of microbes in soil that specifically benefit that plant, by increasing the availability of nutrients or by promoting mycorrhizal associations. This may in turn increase increase the relative performance of that species, reinforcing its dominance in the plant community: this is called positive feedback. In contrast, negative feedback occurs when plant species perform worse in their own soil, perhaps due to an accumulation of host-specific pathogens, thereby preventing any single species becoming dominant, enabling more species to coexist. These kinds of feedback mechanisms clearly have important consequences for plant diversity and vegetation change, with positive feedback potentially leading to the loss of species due to competitive dominance and negative feedback acting to increase diversity due to reduced dominance of plant species within the soil community.

Feedbacks between plants and soil communities that operate via chemical means are also of importance for plant competitive interactions in certain situations. For example, an increasing number of studies point to the influence of plant phenolic compounds on soil biological activity and nutrient availability, which ultimately influence plant competitive interactions (Hättenschwiler and Vitousek 2000). Also, studies reveal the importance of resource partitioning between soil and plant communities, especially in relation to the seasonal and spatial partitioning of limiting resources such as N. As shown in high mountain ecosystems, strong seasonal shifts in soil microbial community structure occur that relate to the temporal partitioning of N between plant and microbial pools (Schadt et al. 2003). Also, there is evidence from strongly N-limited tundra ecosystems that plants, and perhaps microbial communities, partition N on the basis of chemical form

(McKane et al. 2002). All these studies provide exciting insights into the strong, mutual relationships that exist between plant and soil communities that was until recently unrecognized.

There is accumulating evidence that feedback between plant and soil communities is also instrumental in determining the way that ecosystems respond to external forces, such as herbivory and also global change phenomena. Herbivores, which are omnipresent in terrestrial ecosystems, instigate a range of feedbacks between plants and soil biota that determine their impact on ecosystems and their ability to support herbivore populations. For example, they can influence soil biological properties by changing the quality and quantity of resources that enter the soil. This can be via short-term physiological responses of plants to herbivory at the individual plant level (e.g. enhanced root exudation) or longer-term changes in plant productivity and vegetation composition, and subsequent changes in the quantity and quality of litter inputs to soil. In turn, effects of herbivores on soil biological properties feedback above-ground with both positive and negative consequences for primary productivity and herbivore carrying capacity, depending on context and the intensity of different mechanisms that operate. It is also becoming recognized that plant-soil feedbacks are instrumental in determining the ecosystem level consequences of many global change phenomena, such as climate change and N deposition. Numerous studies show how the effects of global change on ecosystems depend on the interactions between plants and soil biota, which are often specific to particular ecosystems. For example, effects of climate change on plant production and C sequestration appear to depend on the extent that the microbial community is nutrient limited (Hu et al. 2001). Also, the impact of N deposition on ecosystem structure depends, in part, on the ability of microbes to sequester added inputs and transfer this N to stable organic matter pools (Zogg et al. 2000); this, in turn, depends on a range of factors such as microbial community composition and constraints on microbial growth, which vary in space and time. In sum, to understand the consequences of global change phenomena for ecosystems requires explicit consideration of linkages between above-ground and below-ground biota.

There are clearly many research challenges that face those interested in soil biology in the future. In particular, there is a need for a stronger theoretical base for soil biology, which in turn requires a deeper mechanistic understanding of the causes and consequences of soil biodiversity, and of the nature and significance of feedback mechanisms that operate between plants and soil biota. As noted above, a particular theme that is emerging is that the consequences for above-ground communities of below-ground interactions, and *vice versa*, are not easily predicted, in that they have both positive and negative outcomes. A key challenge, therefore, is to unravel the nature of this context dependency, which is most likely related to the spatial and temporal scale that interactions operate at, and to abiotic factors that interact with biotic interactions to drive community and ecosystem properties. This presents a major challenge that can only be tackled effectively if soil ecologists work with scientists from other disciplines, such as plant ecologists, molecular biologists, soil physicists and biogeochemists, and theoreticians. The future, therefore, relies on interdisciplinary ecology and explicit recognition of the role of soils and their biota as regulatory forces in Earth's ever-changing ecosystems.

In closing this book, I refer the reader back to its starting words: *few things matter more to human communities than their relations with the soil.* If there is one thing that this book hopefully reveals, it is that the biology of soil is of fundamental importance to the sustainability of life on Earth. Not only will new insights into the biology of soil improve our understanding of how managed and natural ecosystems are structured and function, it will better enable humans to predict the effects on ecosystems of human-induced global changes and enhance our ability to restore degraded ecosystems. Despite this, soil remains the least understood, and perhaps most abused, habitat on Earth.

Bibliography

Abbasi, M.K. and Adams, W.A. (2000). Gaseous N emission during simultaneous nitrification-denitrification associated with mineral fertilization to a grassland soil under field conditions. *Soil Biology and Biochemistry* 32, 1251–1259.

Abe, T., Bignell, D.E., and Higashi, M. (eds.) (2000). *Termites: Evolution, Sociality, Symbioses, Ecology* Kluwer Academic Publishers, Dordrecht, Netherlands.

Aber, J., Nadelhoffer, K., Steudler, P., and Melillo, J.M. (1989). Nitrogen saturation in northern forest ecosystems. *BioScience* 39, 378–386.

Aber, J., McDowell, W., Nadelhoffer, K., Magill, A., Berntson, G., Kamakea, M., et al. (1998). Nitrogen saturation in temperate forests. *BioScience* 48, 921–934.

Addington, R.N. and Seastedt, T.R. (1999). Activity of soil microarthropods beneath snowpack in Alpine tundra and subalpine forest. *Pedobiologia* 43, 47–53.

Aerts, R. (1997). Nitrogen partitioning between resorption and decomposition pathways: a trade off between nutrient use efficiency and litter decomposability? *Oikos* 80, 603–606.

Aerts, R. and Berendse, F. (1988). The effect of increased nutrient availability on vegetation dynamics in wet heathland. *Vegitatio* 76, 63–69.

Alban, D.H. and Berry, E. (1994). Effects of earthworm invasion on morphology, carbon, and nitrogen of a forest soil. *Applied Soil Ecology* 1, 246–249.

Allen, A.S., Andrews, J.A., Finzi, A.C., Matamala, R., Richter, D.D., and Schlesinger, W.H. (2000). Effects of free-air CO_2 enrichment (FACE) on belowground processes in a *Pinus taeda* forest. *Ecological Applications* 10, 437–448.

Allen, M.F. (1991). *The Ecology and Mycorrhizae.* Cambridge University Press, Cambridge.

Al-Mufti, M.M., Sydes, C.L., Furness, S.B., Grime, J.P., and Bond, S.R. (1977). A quantitative analysis of shoot phenology and dominance in herbaceous vegetation. *Journal of Ecology* 65, 759–791.

Alphei, J., Bonkowski, M., and Scheu, S. (1996). Protozoa, Nematoda and Lumbricidae in the rhizosphere of *Hordelymus europaeus* (Poaceae): faunal interaction, response of microorganisms and effects on plant growth. *Oecologia* 106, 111–126.

Altieri, M.G. (1991). How can we best use biodiversity in agroecosystems? *Outlook in Agriculture* 20, 15–23.

Anderson, J.M. (1975). The enigma of soil animal species diversity. In: *Progress in Soil Zoology* (ed. J. Vanek), Academia, Prague, pp. 51–58.

Anderson, J.M. (1978). Inter- and intra-habitat relationships between woodland Cryptostigmata species diversity and the diversity of soil and litter microhabitats. *Oecologia* 32, 341–348.

Anderson, J.B. and Kohn, L.M. (1998). Genotyping gene genealogies and genomics bring fungal population genetics above ground. *Trends in Ecology and Evolution* 13, 444–449.

Anderson, R.V., Coleman, D.C., Cole, C.V., and Elliott, E.T. (1981). Effect of the nematodes *Acrobeloides* sp. and *Mesodiplogaster lheritieri* on substrate utilization and nitrogen and phosphorus mineralization in soil. *Ecology* 62, 549–555.

Anderson, J.M., Ineson, P., and Huish, S.A. (1983). Nitrogen and cation mobilization by soil fauna feeding on leaf litter and soil organic matter from deciduous woodlands. *Soil Biology and Biochemistry* 15, 463–467.

Andrén, O., Bengtsson, J., and Clarholm, M. (1995). Biodiversity and species redundancy among litter decomposers. In: *The Significance and Regulation of Soil Biodiversity* (ed. H.P. Collins), pp. 141–151. Kluwer Academic Publisher, Dordrecht.

Augustine, D.J. and Frank, D.A. (2001). Effects of migratory ungulates on spatial heterogeneity of soil nitrogen properties on a grassland ecosystem. *Ecology* 82, 3149–3162.

Augustine, D.J. and McNaughton, S.J. (1998). Ungulate effects on the functional species composition of plant communities: herbivore selectivity and plant tolerance. *Journal of Wildlife Management* 62, 1165–1183.

Avery, B.W. (1980). Soil Classification for England and Wales. Soil Survey Technical Monograph No 14. Rothamsted Experimental Station, Harpenden.

Ayres, E., Heath, J., Possell, M., Black, H.I.J., Kerstiens, G., and Bardgett, R.D. (2004). Tree physiological responses to above-ground herbivory directly modify below-ground processes of soil carbon and nitrogen cycling. *Ecology Letters* 7, 469–479.

Baldwin, I.T., Olson, R.K., and Reiners, W.A. (1983). Protein binding phenolics and the inhibition of nitrification in subalpine balsam fir soils. *Soil Biology and Biochemistry* 15, 419–423.

Bale, J.S., Hodkinson I.D., Block, W., Webb, N.R., Coulson, S.C., and Strathdee, A.T. (1997). Life strategies of arctic terrestrial arthropods. In: *Ecology of Arctic Environments*, (eds. S.J. Woodin and M. Marquiss) Blackwell Scientific Press. pp. 137–165, Oxford, UK.

Bardgett, R.D. (2000). Patterns of below-ground primary succession at Glacier Bay, south-east Alaska. *Bulletin of the British Ecological Society* 31, 40–42.

Bardgett R.D. and Griffiths, B. (1997). Ecology and biology of soil protozoa, nematodes and microarthropods. In: *Modern Soil Microbiology* (eds. J.D. van Elasas, E. Wellington, and J.T. Trevors), Marcell Dekker, NY, pp. 129–163.

Bardgett, R.D. and Cook, R. (1998). Functional aspects of soil animal diversity in agricultural grasslands. *Applied Soil Ecology* 10, 263–276.

Bardgett, R.D. and Chan, K.F. (1999). Experimental evidence that soil fauna enhance nutrient mineralization and plant nutrient uptake in montane grassland ecosystems. *Soil Biology and Biochemistry* 31, 1007–1014.

Bardgett, R.D. and McAlister, E. (1999). The measurement of soil fungal:bacterial biomass ratios as an indicator of ecosystem self-regulation in temperate grasslands. *Biology and Fertility of Soils* 19, 282–290.

Bardgett, R.D. and Shine, A. (1999). Linkages between plant litter diversity, soil microbial biomass and ecosystem function in temperate grasslands. *Soil Biology and Biochemistry* 31, 317–321.

Bardgett, R.D. and Wardle, D.A. (2003). Herbivore mediated linkages between aboveground and belowground communities. *Ecology* 84, 2258–2268.

Bardgett, R.D. and Walker, L.R. (2004). Impact of coloniser plant species on the development of decomposer microbial communities following deglaciation. *Soil Biology and Biochemistry* 36, 555–559.

Bardgett, R.D., Frankland, J.C., and Whittaker, J.B. (1993a). The effects of agricultural practices on the soil biota of some upland grasslands. *Agriculture, Ecosystems and Environment* 45, 25–45.

Bardgett, R.D., Whittaker, J.B., and Frankland, J.C. (1993b). The diet and food preferences of *Onychiurus procampatus* (Collembola) from upland grassland soils. *Biology and Fertility of Soils* 16, 296–298.

Bardgett, R.D., Whittaker, J.B., and Frankland, J.C. (1993c). The effect of collembolan grazing on fungal activity in differently managed upland pastures—a microcosm study. *Biology and Fertility of Soils* 16, 255–262.

Bardgett, R.D., Speir, T.W., Ross, D.J., Yeates, G.W., and Kettles, H.A. (1994). Impact of pasture contamination by copper, chromium and arsenic timber preservative on soil microbial properties and nematodes. *Biology and Fertility of Soils* 18, 71–79.

Bardgett, R.D., Hobbs, P.J., and Frostegård, Å. (1996). Changes in fungal: bacterial biomass ratios following reductions in the intensity of management on an upland grassland. *Biology and Fertility of Soils* 22, 261–264.

Bardgett, R.D., Leemans, D.K., Cook, R., and Hobbs, P.J. (1997). Seasonality of soil biota of grazed and ungrazed hill grasslands. *Soil Biology and Biochemistry* 29, 1285–1294.

Bardgett, R.D., Wardle, D.A., and Yeates, G.W. (1998). Linking above-ground and below-ground food webs: how plant responses to foliar herbivory influence soil organisms. *Soil Biology and Biochemistry* 30, 1867–1878.

Bardgett, R.D., Denton, C.S., and Cook, R. (1999a). Below-ground herbivory promotes soil nutrient transfer and root growth in grassland. *Ecology Letters* 2, 357–360.

Bardgett, R.D., Kandeler, E., Tscherko, D., Hobbs, P.J., Jones, T.H., Thompson, L.J., et al. (1999b). Below-ground microbial community development in a high temperature world. *Oikos* 85, 193–203.

Bardgett, R.D., Mawdsley, J.L., Edwards, S., Hobbs, P.J., Rodwell, J.S., and Davies, W.J. (1999c). Plant species and nitrogen effects on soil biological properties of temperate upland grasslands. *Functional Ecology* 13, 650–660.

Bardgett, R.D., Anderson, J.M., Behan-Pelletier, B., Brussaard, L., Coleman, D.C., Ettema, C., et al. (2001a). The role of soil biodiversity in the transfer of materials between terrestrial and aquatic systems. *Ecosystems* 4, 421–429.

Bardgett, R.D., Jones, A.C., Jones, D.L., Kemmitt, S.J., Cook, R., and Hobbs, P.J. (2001b). Soil microbial community patterns related to the history and intensity of grazing in sub-montane ecosystems. *Soil Biology and Biochemistry* 33, 1653–1664.

Bardgett, R.D., Streeter, T.C., Cole, L., and Hartley, I.R. (2002). Linkages between soil biota, nitrogen availability, and plant nitrogen uptake in a mountain ecosystem in the Scottish Highlands. *Applied Soil Ecology* 19, 121–134.

Bardgett, R.D., Streeter, T., and Bol, R. (2003). Soil microbes compete effectively with plants for organic nitrogen inputs to temperate grasslands. *Ecology* 84, 1277–1287.

Bardgett, R.D., Yeates, G.W., and Anderson, J.M. (2005). Patterns and determinants of soil biological diversity. In: *Biological Diversity and Function in Soil* (eds. R.D. Bardgett, M.B. Usher, and D.W. Hopkins), Cambridge University Press, Cambridge, UK.

Barker, G.M. and Mayhill, P.C. (1999). Patterns of diversity and habitat relationships in terrestrial mollusc communities of the Pukemaru Ecological District, northeastern New Zealand. *Journal of Biogeography* 26, 215–238.

Beare, M.H. (1997). Fungal and bacterial pathways of organic matter decomposition and nitrogen mineralization in arable soils. In: *Soil Ecology in Sustainable Agricultural Systems*, (eds. L. Brussaard and R. Ferrera-Cerrato) CRC Press, Boca Raton, NY, pp. 37–70.

Beare, M.H., Coleman, D.C., Crossley, D.A., Hendrix, P.F., and Odum, E.P. (1995). A hierarchical approach to evaluating the significance of soil biodiversity to biogeochemical cycling. *Plant and Soil.* 170, 5–22

Beare, M.H., Hu, S., Coleman, D.C., and Hendrix, P.F. (1997*a*). Influences of mycelial fungi on soil aggregration and soil organic matter retention in conventional and no-tillage soils. *Applied Soil Ecology* 5, 211–219.

Beare, M.H., Vikram Reddy, M., Tian, G., and Srivastava, S.C. (1997*b*). Agricultural intensification, soil biodiversity and agroecosystem function in the tropics: the role of decomposer biota. *Applied Soil Ecology* 6, 87–108.

Beedlow, P.A., Tingey, D.T., Phillips, D.L., Hogsett, W., and Olszyk, D.M. (2004). Rising atmospheric CO_2 and carbon sequestration in forests. *Frontiers in Ecology and the Environment* 2, 315–322.

Behan-Pelletier, V.M. (1978). *Diversity, Distribution and Feeding Habits of North American Arctic Soil Acari.* Unpublished Ph.D. thesis, McGill University, Montreal. 428 pp.

Behan-Pelletier, V. and Newton, G. (1999). Linking soil biodiversity and ecosystem function—the taxonomic dilemma. *BioScience* 49, 149–153.

Belnap, J. (2002). Nitrogen fixation in biological soil crusts from southeast Utah, USA. *Biology and Fertility of Soils* 35, 128–135.

Belnap, J. (2003). The world at your feet: desert biological soil crusts. *Frontiers in Ecology and the Environment* 1, 181–189.

Belovsky, G.E. and Slade, J.B. (2000). Insect herbivory accelerates nutrient cycling and increases plant production. *Proceedings of the National Academy of Sciences USA* 97, 14412–14417.

Bengtsson, G. and Rundgren, S. (1983). Respiration and growth of a fungus *Mortierella isabellina*, in response to grazing by *Onychiurus armatus* (Collembola). *Soil Biology and Biochemistry* 15, 469–473.

Bengtsson, G., Erlandsson, A., and Rundgren, S. (1988). Fungal odour attracts soil Collembola. *Soil Biology and Biochemistry* 20, 25–30.

Berg, B. and Matzner, E. (1997). Effect of N deposition on decomposition of plant litter and soil organic matter in forest systems. *Environmental Reviews* 5, 1–25.

Berntson, G.M. and Bazzaz, F.A. (1996). Belowground positive and negative feedbacks on CO_2 growth enhancement. *Plant and Soil* 187, 119–131.

Berntson, G.M. and Bazzaz, F.A. (1997). Nitrogen cycling in microcosms of yellow birch exposed to elevated CO_2: simultaneous positive and negative below-ground feedbacks. *Global Change Biology* 3, 247–258.

Bever, J.D. (1994). Feedback between plants and their soil communities in an old field community. *Ecology* 75, 1965–1977.

Bever, J.D. (2002). Negative feedback within a mutualism: host-specific growth of mycorrhizal fungi reduces plant benefit. *Proceedings of the Royal Society of London* 269, 2595–2601.

Bever, J.D. (2003). Soil community feedback and the coexistence of competitors: conceptual frameworks and empirical tests. *New Phytologist* 157, 465–473.

Bever, J.D., Westover, K.M., and Antonovics, J. (1997). Incorporation of the soil community into plant population dynamics: the utility of the feedback approach. *Journal of Ecology* 85, 561–573.

Bezemer, T.M., Wagenaar, R., Van Dam, N.M., and Wäckers, F.L. (2004). Interactions between above- and belowground insect herbivores as mediated by the plant defense system. *Oikos* 101, 555–562.

Billes, G., Rouhier, H., and Bottner, P. (1993). Modifications of the carbon and nitrogen allocations in the plant (*Triticum-aestivum* L) soil system in response to increased atmospheric CO_2 concentration. *Plant and Soil* 157, 215–225.

Billings, W.D. (1987). Carbon balance of Alaskan tundra and taiga ecosystems: past, present, and future. *Quaternary Science Reviews* 6, 165–177.

Billings, S.A., Schaeffer, S.M. and Evans, R.D. (2002*a*). Trace N gas losses and N mineralization in Mojave desert soils exposed to elevated CO_2. *Soil Biology and Biochemistry* 34, 1777–1784.

Billings, S.A., Schaeffer, S.M., Zitzer, S., Charlet, T., Smith, S.D., and Evans, R.D. (2002*b*). Alterations of nitrogen dynamics under elevated carbon dioxide in an intact Mojave Desert ecosystem: evidence from nitrogen-15 natural abundance. *Oecologia* 131, 463–467.

Blackshaw, R.P. (1990). Studies on *Artioposthia triangulata* (Dendy) (Tricladida: Terricola), a predator of earthworms. *Annals of Applied Biology* 116, 169–176.

Blair, J.M., Parkinson, D., and Parsons, W.F.J. (1990). Decay rates, nitrogen fluxes and the decomposer communities of single and mixed species foliar litters. *Ecology* 71, 1976–1985.

Bloemers, G.F., Hodda, M., Lambshead, P.J.D., Lawton, J.H., and Wanless, F.R. (1997). The effects of forest disturbance on diversity of tropical soil nematodes. *Oecologia* 111, 575–582.

Boag, B. (2000). The impact of the New Zealand flatworm on earthworms and moles in agricultural land in western Scotland. *Aspects of Applied Biology* 62, 79–84.

Boag, B. and Yeates, G.W. (1998). Soil nematode biodiversity in terrestrial ecosystems. *Biodiversity and Conservation* 7, 617–630.

Boag, B. and Yeates, G.W. (2001). The potential impact of the New Zealand flatworm, a predator of earthworms, in western Europe. *Ecological Applications* 11, 1276–1286.

Boag, B., Evans, K.A., Neilson, R., Yeates, G.W., Johns, P.M., Mather, J.G., et al. (1995). The potential spread of terrestrial planarians *Artioposthia triangulata* and *Australoplana sanguinea* var. *alba* to continental Europe. *Annals of Applied Biology* 127, 385–390.

Boag, B., Jones, H.D., and Neilson, R. (1997). The spread of the New Zealand earthworm within Great Britain. *European Journal of Soil Biology* 33, 53–56.

Boddy, L. (1999). Saprotrophic cord-forming fungi: meeting the challenge of heterogeneous environments. *Mycologia* 91, 13–32.

Bohlen, P. (2002). Earthworms. In: *Encyclopaedia of Soil Science*, (ed. R. Lall), Marcell Dekker, New York, pp. 370–373.

Bohlen, P.J., Groffman, P.M., Fahey, T.J., Fisk, M.C., Suárez, E., Pelletier, D.M., et al. (2004*a*). Ecosystem consequences of exotic earthworm invasion of north temperate forests. *Ecosystems* 7, 1–12.

Bohlen, P.J., Pelletier, D.M., Groffman, P.M., Fahey, T.J., and Fisk, M.C. (2004*b*). Influence of earthworm invasion on redistribution and retention of soil carbon and nitrogen in northern temperate forests. *Ecosystems* 7, 13–27.

Bohlen, P.J., Scheu, S., Hale, C.M., McLean, M.A., Migge, S., Groffman, P.M., et al. (2004*c*). Non-native invasive earthworms as agents of change in northern temperate forests. *Frontiers in Ecology and the Environment* 2, 427–435.

Bokhari, U.G. (1977). Regrowth of western wheatgrass utilizing 14C [carbon isotope]-labeled assimilates stored in belowground parts. *Plant and Soil* 48, 115–127.

Bongers, T. and Bongers, M. (1998). Functional diversity of nematodes. *Applied Soil Ecology* 10, 239–251.

Bonkowski, M. (2004). Protozoa and plant growth: the microbial loop revisited. *New Phytologist* 162, 617–631.

Bonkowski, M. and Brandt, F. (2002). Do soil protozoa enhance plant growth by hormonal effects? *Soil Biology and Biochemistry* 34, 1709–1715.

Bonkowski, M., Griffiths, B.S., and Scrimgeour, C. (2000). Substrate heterogeneity and microfauna in soil organic 'hotspots' as determinants of nitrogen capture and growth of ryegrass. *Applied Soil Ecology* 14, 37–53.

Bonkowski, M., Geoghegan, I.E., Birch, A.N.E., and Griffiths, B.S. (2001*a*). Effects of soil decomposer invertebrates (protozoa and earthworms) on an above-ground phytophagous insect (cereal aphid) mediated through changes in the host plant. *Oikos* 95, 441–450.

Bonkowski, M., Jentschke, G., and Scheu, S. (2001*b*). Contrasting effects of microbes in the rhizosphere: interactions of mycorrhiza (*Paxillus involutus* (Batsch) Fr.), naked amoebae (Protozoa) and Norway Spruce seedlings (*Picea abies* Karst.). *Applied Soil Ecology* 18, 193–204.

Boschker, H.T.S., Nold, S.C., Wellsbury, P., Bos, D., de Graaf, W., Pel, R., et al. (1998). Direct linking of microbial populations to specific biogeochemical processes by ^{13}C-labelling of biomarkers. *Nature* 392, 810–815.

Bouwman, L.A., Bloem, J., van den Boogert, P.H.J.F., Bremer, F., Hoenderboom, G.H.J., and de Ruiter, P.C. (1994). Short-term and long-term effects of bacterivorous nematodes and nematophagous fungi on carbon and nitrogen mineralization in microcosms. *Biology and Fertility of Soils* 17, 249–256.

Bowman, W.D. (1992). Inputs and storage of nitrogen in winter snowpack in an alpine ecosystem. *Arctic and Alpine Research* 24, 211–215.

Bowman, W.D. and Steltzer, H. (1998). Positive feedbacks to anthropogenic nitrogen deposition in Rocky Mountain alpine tundra. *Ambio* 27, 514–517.

Bowman, W.D., Steltzer, H., Rosenstiel, T.N., Cleveland, C.C., and Meier, C.L. (2004). Litter effects of two co-occurring alpine species on plant growth, microbial acticity and immobilization of nitrogen. *Oikos* 104, 336–344.

Bradford, M.A., Jones, T.H., Bardgett, R.D., Black, H.I.J., Boag, B., Bonkowski, M., et al. (2002). Impacts of soil faunal community composition on model grassland ecosystems. *Science* 298, 615–618.

Brady, R.L. and Weil, R.R. (1999). *The Nature and Properties of Soils*, 13th edn. Prentice Hall, Upper Saddle River, NJ.

Brand, R.H. and Dunn, C.P. (1998). Diversity and abundance of springtails (Insecta: Collembola) in native and restored tallgrass prairies. *American Midland Naturalist* 139, 235–242.

Bret-Harte, M.S., Shaver, G.R., and Chapin, F.S. (2002). Primary and secondary stem growth in arctic shrubs: implications for community response to environmental change. *Journal of Ecology* 90, 251–267.

Briones, M.J.I., Ineson, P., and Piearce, T.G. (1997). Effects of climate change on soil fauna; responses of enchytraeids, Diptera larvae and tardigrades in a transplant experiment. *Applied Soil Ecology* 6, 117–134.

Briones, M.J.I., Ineson, P., and Poskitt, J. (1998). Climate change and *Cognettia sphagnetorum*: effects on carbon dynamics in organic soils. *Functional Ecology* 12, 528–535.

Brooker, R. and Van der Wal, R. (2003). Can soil temperature direct the composition of high Arctic plant communities? *Journal of Vegetation Science* 14, 535–542.

Brooks, M.L. and Esque, T.C. (2002). Alien annual grasses and fire in the Mojave Desert. *Madroñp* 46, 13–19.

Brooks, M.L., D'Antonio, C.M., Richardson, D.M., Grace, J.B., Keeley, J.E., DiTomaso, J.M., et al. (2004). Effects of invasive alien plants on fire regimes. *BioScience* 54, 677–688.

Brooks, P.D., Williams, M.W., and Schmidt, S.K. (1995). Snowpack controls on soil nitrogen dynamics in the Colorado alpine. In: *Biogeochemistry of Snow-Covered Catchments* (eds. K. Tonnessen, M. Williams, and M. Tranter), International Association of Hydrological Sciences Publication 228, IAHS Publications, Wallingford, pp. 283–292.

Brooks, P.D., Williams, M.W., and Schmidt, S.K. (1996). Microbial activity under alpine snow packs, Niwot Ridge, Co. *Biogeochemistry* 32, 93–113.

Brown, G.G. (1995). How do earthworms affect microfloral and faunal community diversity? *Plant and Soil* 170, 209–231.

Brown, V.K. and Gange, A.C. (1990). Insect herbivory belowground. *Advances in Ecological Research* 20, 1–58.

Brown, V.K. and Gange, A.C. (1992). Secondary plant succession: how is it modified by insect herbivory? *Vegetation* 101, 3–13.

Brussaard, L., Bakker, J.P., and Olff, H. (1996). Biodiversity of soil biota and plants in abandoned arable fields and grasslands under restoration management. *Biodiversity and Conservation* 5, 211–221.

Brussaard, L., BehanPelletier, V.M., Bignell, D.E., Brown, V.K., Didden, W., Folgarait, P., et al. (1997). Biodiversity and ecosystem functioning in soil. *Ambio* 26, 563–570.

Bryant, J.P. and Reichart, P.B. (1992). Controls over secondary metabolite production in arctic woody plants. In: *Arctic Ecosystems in a Changing Climate: An Ecophysiological Perspective*, (eds. F.S. Chapin, R.L. Jefferies, J.F. Reynolds, G.R. Shaver, and J. Svoboda), Academic Press, NY, pp. 377–390.

Bryant, J.P., Reichart, P.B., Clausen, T.P., and Werner R.A. (1993). Effects of mineral nutrition on delayed inducible resistance in Alaska paper birch. *Ecology* 74, 2072–2084.

Buckland, S.M. and Grime, J.P. (2000). The effects of trophic structure and soil fertility on the assembly of plant communities: a microcosm experiment. *Oikos* 91, 336–352.

Bull, I.D., Parekh, N.R., Hall, G., Ineson, P., and Evershed, R.P. (2000). Detection and classification of atmospheric methane oxidising bacteria in soil. *Nature* 405, 175–178.

Burton, A.J., Pregitzer, K.S., Crawford, J.N., Zogg, G.P., and Zak, D.R. (2004). Simulated chronic NO_3^- deposition reduces soil respiration in northern hardwood forest. *Global Change Biology* 10, 1080–1091.

Carreiro, M.M., Sinsabaugh, R.L., Repert, D.A., and Pankhurst, D.F. (2000). Microbial enzyme shifts explain litter decay responses to simulated nitrogen deposition. *Ecology* 81, 2359–2365.

Carroll, J.A., Johnson, D., Morecroft, M., Taylor, A., Caporn, S.J.M., and Lee, J.A. (2000). The effect of long-term nitrogen additions on the bryophyte cover of upland acidic grasslands. *Journal of Bryology* 22, 83–89.

Cates, R.G. and Orians, G.H. (1975). Successional status and the palatability of plants to generalized herbivores. *Ecology* 56, 410–418.

Cerri, C.C., Bernous, M., Cerri, C.E.P, and Feller, C. (2004). Carbon cycling and sequestration opportunities in South America: the case of Brazil. *Soil Use and Management* 20, 248–254.

Chalupský, J. (1995). Long-term study of Enchytraeidae (Oligochaeta) in man-impacted mountain forest soils in the Czech Republic. *Acta Zoologica Fennica* 196, 318–320.

Chan, K.Y. (2001). An overview of some tillage impacts on earthworm population abundance and diversity—implications for functioning in soils. *Soil and Tillage Research* 57, 179–191.

Chapin, F.S. (1980). Mineral nutrition of wild plants. *Annual Review of Ecology and Systematics* 11, 233–260.

Chapin, F.S. (1991). Integrated responses of plants to stress. *BioScience* 41, 29–36.

Chapin, F.S., Moilanen, L., and Kielland, K. (1993). Preferential use of organic nitrogen for growth by a non-mycorrhizal arctic sedge. *Nature* 361, 150–153.

Chapin, F.S., Walker, L.R., Fastie, C.L., and Sharman. L.C. (1994). Mechanisms of primary succession following deglaciation at Glacier Bay, Alaska. *Ecological Monographs* 64, 149–175.

Chapin, F.S., Bert-Harte, S.M., Hobbie, S.E., and Zhong, H. (1996a). Plant functional types as predictors of transient responses of arctic vegetation to global change. *Journal of Vegetation Science* 7, 347–358.

Chapin, F.S., Reynolds, H., D'Antonio, C.M., and Eckhart, V. (1996b). The functional role of species in terrestrial ecosystems. In: *Global Change in Terrestrial Ecosystems* (eds. B. Walker and W. Steffan), Cambridge University Press, Cambridge, pp. 403–430.

Chapin, F.S., Matson, P.A., and Mooney, H.A. (2002). *Principles of Terrestrial Ecosystem Ecology.* Springer, NY.

Chapman, K., Whittaker, J.B., and Heal, O.W. (1988). Metabolic and faunal activity in litters of tree mixtures compared with pure stands. *Agriculture Ecosystems and Environment* 24, 33–40.

Chapman, W.L. and Walsh, J.E. (1993). Recent variations of sea ice and air temperature in high latitudes. *Bulletin American Meteorological Society* 74, 33–47.

Chauvel, A., Grimaldi, M., Barros, E., Blanchart, E., Desjardins, T., Sarrazin, M., and Lavelle, P. (1999). Pasture damage by an Amazonian earthworm. *Nature* 398, 32–33.

Cho, J.C. and Tiedje, J.M. (2000). Biogeography and degree of endemicity of fluorescent *Pseudomonas* strains in soil. *Applied and Environmental Microbiology* 66, 5448–5456.

Clarholm, M. (1985). Interactions of bacteria, protozoa and plants leading to mineralization of soil nitrogen. *Soil Biology and Biochemistry* 17, 181–187.

Clarholm, M. (1994). The microbial loop. In: *Beyond the Biomass* (eds. K. Ritz, J. Dighton, and K.E. Giller), Blackwell Scientific Publications, Oxford, UK, pp. 355–365.

Cleveland, C.C., Townsend, A.R., Schimel, D.S., Fisher, H., Howarth, R.W., Hedin, L.O., et al. (1999). Global patterns of terrestrial nitrogen (N_2) fixation in natural ecosystems. *Global Biogeochemical Cycles* 13, 623–645.

Cleveland, C.C., Townsend, A.R., and Schmidt, S.K. (2002). Phosphorus limitation of microbial processes in moist tropical forests: evidence from short-term laboratory incubations and field studies. *Ecosystems* 5, 680–691.

Cole, L., Bardgett, R.D., and Ineson, P. (2000). Enchytraeid worms (Oligochaeta) enhance carbon mineralization in organic upland soils. *European Journal of Soil Science* 51, 185–192.

Cole, L., Bardgett, R.D., Ineson, P., and Adamson, J. (2002*a*). Relationships between enchytraeid worms (Oligochaeta), temperature, and the release of dissolved organic carbon from blanket peat in northern England. *Soil Biology and Biochemistry* 34, 599–607.

Cole, L., Bardgett, R.D., Ineson, P., and Hobbs, P.J. (2002*b*). Enchytraeid worm (Oligochaeta) influences on microbial community structure, nutrient dynamics, and plant growth in blanket peat subjected to warming. *Soil Biology and Biochemistry* 34, 83–92.

Cole, L.C., Staddon, P.L., Sleep, D., and Bardgett, R.D. (2004). Soil animals influence microbial abundance, but not plant-microbial competition for soil organic. *Functional Ecology* 18, 631–640.

Coleman, D.C., Reid, C.P.P., and Cole, C.V. (1983). Biological strategies of nutrient cycling in soil systems. *Advances in Ecological Research* 13, 1–55.

Connell, J.H. (1978). Diversity in tropical rainforests and coral reefs. *Science* 199, 1302–1310.

Connell, J.H. and Lowman, M.D. (1989). Low-diversity tropical rain forests: some possible mechanisms for their existence. *American Naturalist* 134, 88–119.

Coomes, D.A., Allen, R.B., Forsyth, D.M., and Lee, W.G.(2003). Factors preventing the recovery of New Zealand forests following control of invasive deer. *Conservation Biology* 17, 450–459.

Cornelissen, J.H.C. and Thompson, K. (1997). Functional leaf attributes predict litter decomposition rate in herbaceous plants. *New Phytologist* 135, 573–582.

Cornelissen, J.H.C., Perez-Harguindeguy, N., Diaz, S., Grime, J.P., Marzano, B., Cabido, M., et al. (1999). Leaf structure and defence control litter decomposition rate across species, life forms and continents. *New Phytologist* 143, 191–200.

Cortez, J., Hameed, R., and Bouché, M.B. (1989). C and N transfer in soil with or without earthworms fed with ^{14}C- and ^{15}N labelled wheat straw. *Soil Biology and Biochemistry* 21, 491–497.

Coulson, S.J., Leinass, H.P., Ims, R.A., and Søvik, G. (2000). Experimental manipulation of the winter surface ice layer: the effects on Arctic soil microarthropod community. *Ecography* 23, 299–306.

Coulson, S.J., Hodkinson, I.D., and Webb, N.R. (2003). Microscale distribution patterns in high Arctic soil microarthropod communities: the influence of plants species within the vegetation mosaic. *Ecography* 26, 801–809.

Courtright, E.M., Wall, D.H., and Virginia, R.A. (2001). Determining habitat suitability for soil invertebrates in an extreme environment: the McMurdo Dry Valleys, Antarctica. *Antarctic Science* 13, 9–17.

Coûteaux, M.M., Kurz, C., Bottner, P., and Raschi, A. (1999). Influence of increased atmospheric CO_2 concentration on quality of plant material and litter decomposition. *Tree Physiology* 19, 301–311.

Cragg, R.H. and Bardgett, R.D. (2001). How changes in animal diversity within a soil trophic group influence ecosystem processes. *Soil Biology and Biochemistry* 33, 2073–2081.

Crawley, M.J. (1983). *Herbivory: The Dynamics of Animal–Plant Interactions.* Blackwell Scientific Publications, Oxford.

Crawley, M.J. (1997). Plant–herbivore interactions. In: *Plant Ecology* (ed. M.J. Crawley), Blackwell Scientific Publications, Oxford, pp. 401–474.

Crews, T.E., Kitayama, K., Fownes, J.H., Riley, D.A., Herbert, D.A., Mueller-Dombois, D., et al. (1995). Changes in soil phosphorus fractions and ecosystem dynamics across a long chronosequence in Hawaii. *Ecology* 76, 1407–1424.

Critchley, B.R., Cook, A.G., Critchley, U., Perfect, T.J., Russell-Smith, A., and Yeadon, R. (1979). Effects of bush clearing and soil cultivation on the invertebrate fauna of a forest soil in the humid tropics. *Pedobiologia* 19, 425–438.

Crocker, R.L. and Major, J. (1955). Soil development in relation to vegetation and surface age at Glacier Bay, Alaska. *Journal of Ecology* 43, 427–448.

Culter, D.W., Clump, L.M., and Sandon, H. (1923). A quantitative investigation of the bacterial and protozoan populations of the soil. *Philosophical Transactions of the Royal Society, London B* 211, 317–350.

Curtis, P.S. and Wang, X.Z. (1998). A meta-analysis of elevated CO_2 effects on woody plant mass, form, and physiology. *Oecologia* 113, 299–313.

Curtis, T.P., Sloan, W.T., and Scannell, J.W. (2002). Estimating prokaryotic diversity and its limits. *Proceedings of the National Academy of Sciences USA*, 99, 10494–10499.

D'Antonio, C. and Vitousek, P.M. (1992). Biological invasions by exotic grasses, the grass/fire cycle and global change. *Annual Review of Ecology and Systematics* 23, 63–87.

D'Antonio, C., Meyerson, L.A., and Denslow, J. (2001). Exotic species and conservation. In: *Conservation Biology: Research Priorities for the New Decade* (eds. M.E. Soule and G.H. Orians), Island Press, Washington D.C, pp. 59–80.

Daniell, T.J., Husband, R., Fitter, A.H., and Young, J.P.W. (2001). Molecular diversity of arbuscular mycorrhizal fungi colonising arable crops. *FEMS Microbiology Ecology* 36, 203–209.

Darwin, C. (1881). *The Formation of Vegetable Mould, through the Action of Worms, with Observations on Their Habits.* John Murray, London.

De Deyn, G.B., Raaijmakers, C.E., Zoomer, H.R., Berg, M.P., De Ruiter, P.C., Verhoef, H.A., et al. (2003). Soil invertebrate fauna enhances grassland succession and diversity. *Nature* 422, 711–713.

De Deyn, G.B., Raaijmakers, C.E., Van Ruijven, J., Berendse, F., and Van der Putten, W.H. (2004). Plant species identity and diversity effects on different trophic levels of nematodes in the soil food web. *Oikos* 106, 576–586.

DeLuca, T.H., Zackrisson, O., Nilsson, M.C., and Sellstedt, A. (2002). Quantifying nitrogen fixation in feather moss carpets of boreal forest. *Nature* 419, 917–920.

De Mazancourt, C., Loreau, M., and Abbadie, L. (1999). Grazing optimization and nutrient cycling: potential impact of large herbivores in a savannah system. *Ecological Applications* 9, 784–797.

Denton, C.S., Bardgett, R.D., Cook, R., and Hobbs, P.J. (1999). Low amounts of root herbivory positively influences the rhizosphere microbial community of a temperate grassland soil. *Soil Biology and Biochemistry* 31, 155–165.

De Ruiter, P.C., Neutel, A.N., and Moore, J.C. (1995). Energetics, patterns of interaction strengths, and stability in real ecosystems. *Science* 269: 1257–1260.

Dhillion, S.S., Roy, J., and Abrams, M. (1996). Assessing the impact of elevated CO_2 on soil microbial activity in a Mediterranean model ecosystem. *Plant and Soil* 187, 333–342.

Díaz, S., Grime, J.P., Harris, J., and McPherson, E. (1993). Evidence of a feedback mechanism limiting plant-response to elevated carbon-dioxide. *Nature* 364, 616–617.

Díaz, S., Cabido, M., and Casanoves, F. (1998). Plant functional types and environmental filters at a regional scale. *Journal of Vegetation Science* 9, 113–122.

Didden, W.A.M. (1993). Ecology of terrestrial Enchytraeidae. *Pedobiologia* 37, 2–29.

Didden, W.A.M. and De Fluiter, R. (1998). Dynamics and stratification of Enchytraeidae in the organic layer of a Scots pine forest. *Biology and Fertility of Soils* 26, 305–312.

Dighton, J. (1978). Effects of synthetic lime aphid honeydew on populations of soil organisms. *Soil Biology and Biochemistry* 10, 369–376.

Dlamini, T.C. and Haynes, R.J. (2004). Influence of agricultural land use on the size and composition of earthworm communities in northern KwaZulu-Natal, South Africa. *Applied Soil Ecology* 27, 77–88.

Dokuchaev, V.V. (1879). Abridged historical account and critical examination of the principal soil classifications existing. *Transactions of the Petersburg Society of Naturalists* 1, 64–67.

Dominy, C.S., Haynes, R.J., and van Antwerpen, R. (2002). Loss of organic matter and related soil properties under long-term sugarcane production on two contrasting soils. *Biology and Fertility of Soils* 36, 350–356.

Donnison, L.M., Griffith, G.S., Hedger, J., Hobbs, P.J., and Bardgett, R.D. (2000a). Management influences on soil microbial communities and their function in botanically diverse haymeadows of northern England and Wales. *Soil Biology and Biochemistry* 32, 253–263.

Donnison, L.M., Griffith, G.S., and Bardgett, R.D. (2000b). Determinants of fungal growth and activity in botanically diverse haymeadows: effects of litter type and fertilizer additions. *Soil Biology and Biochemistry* 32, 289–294.

Drake, J.A. and Mooney, H.A. (1989). *Biological Invasions: A Global Perspective.* Wiley, Chichester, UK.

Edwards, C.A. (2004). *Earthworm Ecology*, 2nd edn. CRC Press, Boca Raton, FL.

Edwards, C.A. and Cresser, M.S. (1992). Freezing and its effect on chemical and biological properties of soil. *Advances in Soil Science* 18, 59–79.

Edwards, C.A. and Bohlen, P.J. (1996). *Biology and Ecology of Earthworms*, 3rd edn. Chapman and Hall, London.

Egerton-Warburton, L.M. and Allen, E.B. (2000). Shifts in arbuscular mycorrhizal communities along an anthropogenic nitrogen deposition gradient. *Ecological Applications* 10, 484–496.

Egerton-Warburton, L.M., Graham, R.C., Allen, E.B., and Allen, M.F. (2001). Reconstruction of the historical changes in mycorrhizal fungal communities under anthropogenic nitrogen deposition. *Proceedings of The Royal Society of London Series B-Biological Sciences* 268, 2479–2484.

Eggleton, P. and Bignell, D.E. (1995). Monitoring the response of tropical insects to changes in the environment: troubles with termites. In: *Insects in a Changing Environment* (eds. R. Harrington and N.E. Stork), Academic Press, London, pp. 434–497.

Eggleton, P., Williams, P.H., and Gaston, K.J. (1994). Explaining global termite diversity: productivity or history. *Biodiversity and Conservation* 3, 318–330.

Eggleton, P., Bignell, D.E., Hauser, S., Dibog, L., Norgrove, L., and Madong, B. (2002). Termite diversity across an anthropogenic disturbance gradient in the humid forest zone of West Africa. *Agriculture, Ecosystems and Environment* 90, 189–202.

Eldridge, D.J. (1993). Effect of ants on sandy soils in semiarid eastern Australia: local distribution of nest entrances and their effect on infiltration of water. *Australian Journal of Soil Research* 31, 509–518.

Epstein, H.E., Walker, M.D., Chapin, F.S., and Starfield, A.M. (2000). A transient, nutrient-based model of acrtic plant community response to climate warming. *Ecological Applications* 10, 824–841.

Erland, S. and Taylor, A.F.S. (2002). Diversity of ectomycorrhizal fungal communities in relation to the abiotic environment. In: *Mycorrhizal Ecology* (eds. G.A. van der Hiedjen and I. Sanders), Springer-Verlag, Berlin Hiedelberg, pp. 163–200.

Ettema, C.H. and Wardle, D.A. (2002). Spatial soil ecology. *Trends in Ecology and Evolution* 17, 177–183.

Ettema, C.H., Lowrance, R., and Coleman, D.C. (1999). Riparian soil response to surface nitrogen input: the indicator potential of free-living soil nematode populations. *Soil Biology and Biochemistry* 31, 1625–1638.

Ettema, C.H., Rathbun, S.L., and Coleman, D.C. (2000). On spatiotemporal patchiness and the coexistence of five species of Chronogaster (Nematoda: Chronogasteridae) in a riparian wetland. *Oecologia* 125, 444–452.

Evans, R.D. and Ehleringer, J.R. (1993). A break in the nitrogen cycle in aridlands? Evidence from $\delta^{15}N$ of soil. *Oecologia* 94, 314–317.

Evans, R.D. and Lange, O.L. (2001). Biological soil crusts and ecosystems nitrogen and carbon dynamics. In: *Biological Soil Crusts: Structure, Function and Management* (eds. J. Belnap and O.L. Lange), Springer-Verlag, Berlin, pp. 263–279.

Farrar, J., Hawes, M., Jones, D., and Lindow, S. (2003). How roots control the flux of carbon to the rhizosphere. *Ecology* 84, 827–837.

Findlay, S., Carreiro, M., Krishik, V., and Jones, C.G. (1996). Effects of damage to living plants on leaf litter quality. *Ecological Applications* 6, 269–275.

Finlay, B.J. (2002). Global dispersal of free-living microbial eukaryote species. *Science* 296, 1061–1063.

Finlay, B.J., Esteban, G.F., Olmo, J.L., and Tyler, P.A. (1999). Global distribution of free living microbial species. *Ecography* 22, 138–144.

Finlay, B.J., Esteban, G.F., Olmo, J.L., and Tyler, P.A. (2001). Biodiversity of terrestrial protozoa appears homogenous across local and regional spatial scales. *Protist* 154, 355–366.

Finzi, A.C., DeLucia, E.H., Hamilton, J.G., Richter, D.D., and Schelsinger, W.H. (2002). The nitrogen budget of a pine forest under free-air CO_2 enrichment. *Oecologia* 132, 567–578.

Fischer, K., Hahn, D., Hönerlage, W., and Zeyer, J. (1997). Effect of passage through the gut of the earthworm *Lumbricus terrestris* L. on *Bacillus megaterium* studied by whole cell hybridization. *Soil Biology and Biochemistry* 29, 1149–1152.

Fisk, M.C. and Schmidt, S.K. (1996). Microbial responses to nitrogen additions in alpine tundra soil. *Soil Biology and Biochemistry* 28, 751–755.

Fisk, M.C., Fahey, T.J., Groffman, P.M., and Bohlen, P.J. (2004). Earthworm invasion, fine-root distributions, and soil respiration in north temperate forests. *Ecosystems* 7, 55–62.

Floate, M.J.S. (1970*a*). Mineralization of nitrogen and phosphorus from organic materials of plant and animal origin and its significance in the nutrient cycle in grazed upland and hill soils. *Journal of the British Grassland Society* 25, 295–302.

Floate, M.J.S. (1970*b*). Decomposition of organic materials from hill soils and pastures. II. Comparative studies on the mineralization of carbon, nitrogen and phosphorus from plant materials and sheep faeces. *Soil Biology and Biochemistry* 2, 173–185.

Fog, K. (1988). The effect of added nitrogen on the rate of decomposition of organic matter. *Biological Reviews of the Cambridge Philosophical Society* 63, 433–462.

Foissner, W. (1997*a*). Global soil ciliate (Protozoa, Ciliophora) diversity: a probability-based approach using large samples collections from Africa, Australia and Antarctica. *Biodiversity and Conservation* 6, 1627–1638.

Foissner, W. (1997*b*). Soil ciliates (Protozoa: Ciliophora) from evergreen rain forests of Australia, South America and Costa Rica: diversity and description of new species. *Biology and Fertility of Soils* 25, 317–339.

Foissner, W. (1999*a*). Notes on the soil ciliate biota (Protozoa, Ciliophora) from the Shimba Hills in Kenya (Africa): diversity and descriptions of three new genera and ten new species. *Biodiversity and Conservation* 8, 319–389.

Foissner, W. (1999*b*). Protist diversity: estimates of the near-imponderable. *Protist* 150, 363–368.

Fontaine, S., Bardoux, G., Abbadie, L., and Mariotti, A. (2004). Carbon input to soil may decrease soil carbon content. *Ecology Letters* 7, 314–320.

Fragoso, C., Brown, G.G., Patrón, J.C., Blanchart, E., Lavelle, P., Pashanasi, B., et al. (1997). Agricultural intensification, soil biodiversity and agroecosystem function in the tropics: the role of earthworms. *Applied Soil Ecology* 6, 17–35.

Fragoso, C., Kanyonyo, J., Moreno, A., Senapati, B.K., Blanchart, E., and Rodriguez, C. (1999*a*). A survey of tropical earthworms: taxonomy, biogeography and environmental plasticity. In: *Earthworm Management in Tropical Agroecosystems* (eds. P. Lavelle, L. Brussaard, and P. Hendrix), ABI Publishing, NY, pp. 1–26.

Fragoso, C., Lavelle, P., Blanchart, E., Senapati, B., Jiménez, J.J., Martínez, M.A., et al. (1999*b*). Earthworm communities of tropical agroecosystems: origin, structure and influence of management practices. In: *Earthworm Management in Tropical Agroecosystems* (eds. P. Lavelle, L. Brussaard, and P. Hendrix), CAB International, Wallingford UK, pp. 27–55.

Frank, D.A. and Groffman, P.M. (1998). Ungulate vs. landscape control of soil C and N processes in grasslands of Yellowstone National Park. *Ecology* 79, 2229–2241.

Frankland, J.C. (1982). Biomass and nutrient cycling by decomposer basidiomycetes. In: *Decomposer Basidiomycetes: Their Biology and Ecology* (eds. J.C. Frankland, J.N. Hedger, and M.J. Swift), Cambridge University Press, Cambridge, pp. 241–261.

Frankland, J.C. (1998). Fungal succession—unravelling the unpredictable. *Mycological Research* 102, 1–15.

Freckman, D.W., Duncan, D.A., and Larson, J.R. (1979). Nematode density and biomass in an annual grassland ecosystem. *Journal of Range Management* 32, 418–422.

Freeman, C., Ostle, N.J., and Kang, H. (2001). An enzyme latch on global carbon store. *Nature* 409, 149.

Freeman, C., Fenner, N., Ostle, N.J., Kang, H., Dorwick, D.J., Reynolds, B., et al. (2004). Export of dissolved organic carbon from peatlands under elevated carbon dioxide levels. *Nature* 430, 195–198.

Frey, S.D., Knorr, M., Parrent, J.L., and Simpson, R.T. (2004). Chronic nitrogen enrichment affects the structure and function of the soil microbial community in temperate hardwood and pine forests. *Forest Ecology and Management* 196, 159–171.

Gadgill, R.L. (1971). The nutritional role of *Lupinus arboreus* in coastal sand dune forestry. I. The potential influence of undamaged lupin plants on nitrogen uptake by *Pinus radiata*. *Plant and Soil* 34, 357–367.

Gange, A.C. and Brown, V.K. (1989). Effects of root herbivory by an insect on a foliar-feeding species, mediated through changes in the host plant. *Oecologia* 81, 38–42.

Gange, A.C. and West, H.M. (1994). Interactions between arbuscular mycorrhizal fungi and foliar-feeding insects in *Plantago lanceolata* L. *New Phytologist*, 128, 79–87.

Gange, A.C., Brown, V.K., and Sinclair, G.S. (1993). Vesicular–arbuscular mycorrhizal fungi: a determinant of plant community structure in early succession. *Functional Ecology* 7, 616–622.

Garbaye, J. and Le Tacon, F. (1982). Influence of mineral fertilization and thinning intensity on the fruit body production of epigeous fungi in an artificial spruce stand (*Picea excelsa* Link) in north-eastern France. *Acta Ecologia Planta* 3, 153–160.

Gardner, I.C., Clelland, D.M., and Scott, A. (1984). Mycorrhizal improvement in non-leguminous nitrogen fixing associations with particular reference to *Hippophae rhannoides* L. *Plant and Soil* 78, 189–199.

Gastine, A., Scherer-Lorenzen, M., and Leadley, P.W. (2003). No consistent effects of plant diversity on root biomass, soil biota and soil abiotic conditions in temperate grassland communities. *Applied Soil Ecology* 24, 101–111.

Gehring, C.A. and Whitham, T.G. (1994). Interactions between aboveground herbivores and the mycorrhizal mutualists of plants. *Trends in Ecology and Evolution* 9, 251–255.

Gerrish, G. and Mueller-Dombois, D. (1980). Behaviour of native and non-native plants in two tropical rainforests on Ohau, Hawaiian Islands. *Phytocoenologia* 8, 237–295.

Gill, R.W. (1969). Soil microarthropod abundance following old field litter manipulation. *Ecology* 50, 805–816.

Gill, R.A., Polley, H.W., Johnson, H.B., Anderson, L.J., Maherali, H., and Jackson, R.B. (2002). Nonlinear grassland responses to past and future atmospheric CO_2. *Nature* 417, 279–282.

Giller, K., Bignell, D., Lavelle, P., Swift, M.J., Barrios, E., Moreira, F. et al. (2005). Soild biodiversity in rapidly changing tropical landscapes: scaling up and scaling down. In: *Biological Diversity and Function in Soil* (eds. R.D. Bardgett, M.B. Usher, and D.W. Hopkins), Cambridge University Press, Cambridge.

Gillison, A.N., Jones, D.T., Susilo, F.-X., and Bignell, D.E. (2003). Vegetation indicates diversity of soil macroinvertebrates: a case study with termites along a land-use intensification gradient in lowland Sumatra. *Organisms, Diversity and Evolution* 3, 111–126.

Gonzalez, G. and Seastedt, T.R. (2001). Soil fauna and plant litter decomposition in tropical and subalpine forests. *Ecology* 82, 955–964.

Gordon, C., Wynn, J.M., and Woodin, S.J. (2001). Impacts of increased nitrogen supply on high Arctic heath: the importance of bryophytes and phosphorus availability. *New Phytologist* 149, 461–471.

Gorham, E. (1991). Northern peatlands: role in the carbon cycle and probable responses to climatic warming. *Ecological Applications* 1, 182–195.

Gorissen, A. and Cotrufo, M.F. (1999). Elevated carbon dioxide effects on nitrogen dynamics in grasses, with emphasis on rhizosphere processes. *Soil Science Society of America Journal* 63, 1695–1702.

Gough, L., Wookey, P.A., and Shaver, G.R. (2002). Dry heath arctic tundra responses to long-term nutrient and light manipulations. *Arctic, Antarctic and Alpine Research* 34, 211–218.

Goulden, M.L., Wofsy, S.C., Harden, J.W., Trumbore, S.E., Crill, P.M., Gower, S., et al. (1998). Sensitivity of boreal forest carbon balance to soil thaw. *Science* 279, 214–217.

Grace, J.B. (1999). The factors controlling species density in herbaceous plant communities: an assessment. *Perspectives in Plant Ecology, Evolution, and Plant Systematics* 2, 1–28.

Graham, M.H., Haynes, R.J., and Meyer, J.H. (2002). Soil organic matter content and quality: effect of fertilizer applications, burning and trash retention on a long-term sugarcane experiment in South Africa. *Soil Biology and Biochemistry* 34, 93–102.

Grant, J.D. (1983). The activities of earthworms and the fates of seeds. In: *Earthworm Ecology: From Darwin to Vermiculture*, (ed. J.E. Satchell), Chapman and Hall, London, pp. 107–122.

Grayston, S.J., Shenquiang, W., Campbell, C.D., and Edwards, A.C. (1998). Selective influence of plant species on microbial diversity in the rhizosphere. *Soil Biology and Biochemistry* 30, 369–378.

Grayston, S, J., Campbell, C.D., Bardgett, R.D., Mawdsley, J.L., Clegg, C.D., Ritz, K., et al. (2004). Assessing shifts in soil microbial community structure across a range of grasslands of differing management intensity using CLPP, PLFA and community DNA techniques. *Applied Soil Ecology* 25, 63–84.

Grayston, S.J, Griffiths, G., Mawdsley, J.L., Campbell, C., and Bardgett, R.D. (2001a). Accounting for variability in soil microbial communities of temperate upland grasslands. *Soil Biology and Biochemistry* 33, 533–551.

Grayston, S.J., Dawson, L.A., Treonis, A.M., Murray, P.J., Ross, J., Reid, E.J., et al. (2001b). Impact of root herbivory by insect larvae on soil microbial communities. *European Journal of Soil Biology* 37, 277–280.

Griffiths, B.S., Welschen, R., Van Arendonk, J.J.C.M., and Lambers, H. (1992). The effect of nitrate–nitrogen on bacteria and bacterial-feeding fauna in the rhizosphere of different grass species. *Oecologia* 91, 253–259.

Griffiths, B.S., Ritz, K., Ebblewhite, N., and Dobson, G. (1999). Soil microbial community structure: effects of substrate loading rates. *Soil Biology and Biochemistry* 31, 145–153.

Griffiths, B.S., Ritz, K., Bardgett, R.D., Cook, R., Christensen, S., Ekelund, F., et al. (2000). Ecosystem response of pasture soil communities to fumigation-induced microbial diversity reductions: an examination of the biodiversity-ecosystem function relationship. *Oikos* 90, 279–294.

Griffiths, B.S., Kuan, H.L., Ritz, K., Glover, L.A., McCaig, A.E., and Fenwick, C. (2004). The relationship between microbial community structure and functional stability, tested experimentally in an upland pasture soil. *Microbial Ecology* 47, 104–113.

Grime, J.P. (1973). Control of species diversity in herbaceous vegetation. *Journal of Environmental Management* 1, 151–167.

Grime, J.P. (1979). *Plant Strategies and Vegetation Processes*. John Wiley, Chichester, UK.

Grime, J.P. (1997). Biodiversity and ecosystem function: the debate deepens. *Science* 277, 1260–1261.

Grime, J.P. (2001). *Plant Strategies, Vegetation Processes and Ecosystem Properties*. John Wiley, Chichester, UK.

Grime, J.P., Mackey, J.M., Hillier, S.H., and Read, D.J. (1987). Floristic diversity in a model system using experimental microcosms. *Nature* 328, 420–422.

Grime, J.P., Cornelissen, J.H.C., Thompson, K., and Hodgson, J.G. (1996). Evidence of a causal connection between anti-herbivore defence and the decomposition rate of leaves. *Oikos* 77, 489–494.

Groffman, P.M. and Bohlen, P.J. (1999). Soil and sediment biodiversity. *BioScience* 49, 139–148.

Groffman, P.M., Bohlen, P.J., Fisk, M.C., and Fahey, T.J. (2004). Exotic earthworm invasion and microbial biomass in temperate forest soils. *Ecosystems* 7, 45–54.

Guitian, R. and Bardgett, R.D. (2000). Plant and soil microbial responses to defoliation in temperate semi-natural grassland. *Plant and Soil* 220, 271–277.

Gundersen, P., Callescen, I., and deVries, W. (1998). Nitrate leaching in forest ecosystems is related to forest floor C/N ratios. *Environmental Pollution* 102, 403–407.

Hagedorn, F., Bucher, J.B., Tarjan, D., Rusert, P., and Bucher-Wallin, I. (2000). Responses of N fluxes and pools to elevated atmospheric CO_2 in model forest ecosystems with acidic and calcareous soils. *Plant and Soil* 224, 273–286.

Hagedorn, F., Landolt, W., Tarjan, D., Egli, P., and Bucher, J.B. (2002). Elevated CO_2 influences nutrient availability in young beech–spruce communities on two soil types. *Oecologia* 132, 109–117.

Hagerman, S.H., Sakakibara, S.M., and Durall, D.M. (2001). The potential for woody understory plants to provide refuge for ectomycorrhizal inoculum at an interior Douglas fir forest after clear-cutting. *Canadian Journal of Forest Research* 31, 711–721.

Hamilton, E.W. and Frank, D.A. (2001). Can plants stimulate soil microbes and their own nutrient supply? Evidence from a grazing tolerant grass. *Ecology* 82, 2397–2402.

Hammel, K.E. (1997). Fungal degradation of lignin. In: *Driven by Nature: Plant Litter Quality and Decomposition* (eds. G. Cadish and K.E. Giller), CAB International, Wallingford, England, pp. 33–45.

Hanley, M.E., Trofmov, S., and Taylor, G. (2004). Species-level effects more important than functional group-level responses to elevated CO_2: evidence from simulated turves. *Functional Ecology* 18, 304–313.

Hanlon, R.D.G. (1981). Influence of grazing by Collembola on the activity of senescent fungal colonies grown on media of different nutrient concentration. *Oikos* 36, 362–367.

Hanlon, R.D.G. and Anderson, J.M. (1979). Effects of Collembola grazing on microbial activity in decomposing leaf litter. *Oecologia* 38, 93–99.

Hanounik, S.B. and Osborne, W.W. (1977). The relationships between population density of *Meloidogyne inconita* and nicotine content of tobacco. *Nematologica* 23, 147–152.

Hansen, R.A. (2000). Effects of habitat complexity and composition on a diverse litter microarthropod assemblage. *Ecology* 81, 1120–1132.

Hansen, R.A. and Coleman, D.C. (1998). Litter complexity and composition are determinants of the diversity and species composition of oribatid mites *(Acari: Oribatida)* in litterbags. *Applied Soil Ecology*, 9, 17–23.

Harley, J.L. (1971). Fungi in ecosystems. *Journal of Applied Ecology* 8, 627–642.

Harris, W.V. (1964). *Termites: Their Recognition and Control.* Longmans, London.

Harrison, A.F. (1987). *Soil Organic Phosphorus. A Review of World Literature.* CAB International, Oxford.

Harrison, K.A. and Bardgett, R.D. (2003). How browsing by red deer impacts on litter decomposition in a native regenerating woodland in the Highlands of Scotland. *Biology and Fertility of Soils* 38, 393–399.

Harrison, K.A. and Bardgett, R.D. (2004). Browsing by red deer negatively impacts on soil nitrogen availability in regenerating woodland. *Soil Biology and Biochemistry* 36, 115–126.

Hart, S.C. and Stark, J.M. (1997). Nitrogen limitation of the microbial biomass in an old growth forest. *Ecoscience* 4, 91–98.

Hartley, S.E. and Jones, C.J. (1997). Plant chemistry and herbivory, or why is the world green? In: *Plant Ecology* (ed. M.J. Crawley), Blackwell Science, Oxford, pp. 284–324.

Hartnett, D.C. and Wilson, G.W.T. (1999). Mycorrhizae influence plant community structure and diversity in tallgrass prairie. *Ecology* 80, 1187–1195.

Hassall, M., Turner, J.G., and Rands, M.R.W. (1987). Effects of terrestrial isopods on the decomposition of woodland leaf litter. *Oecologia* 72, 597–604.

Hättenschwiler, S. and Gasser, P. (2005). Soil animals alter plant litter diversity effects on decomposition. *Proceedings of the National Academy of Science* 102, 1519–1524.

Hättenschwiler, S. and Vitousek, P.M. (2000). The role of polyphenols in terrestrial ecosystem nutrient cycling. *Trends in Ecology and Evolution* 15, 238–243.

Haystead, A., Malajczuk, N., and Grove, T.S. (1988). Underground transfer of nitrogen between pasture plants infected with arbuscular mycorrhizal fungi. *New Phytologist* 108, 417–423.

Hector, A., Beale, A.,J., Minns, A., Otway, S.J., and Lawton, J.H. (2000). Consequences of the reduction of plant litter diversity for decomposition: effects through litter quality and microenvironment. *Oikos* 90, 357–371.

Hedlund, K., Boddy, L., and Preston, C.M. (1991). Mycelial responses of the soil fungus *Mortierella isabellina*, to grazing by *Onychiurus armatus* (Collembola). *Soil Biology and Biochemistry* 23, 361–366.

Heemsbergen, D.A., Berg, M.P., Loreau, M., van Hal, J.R., Faber, J.H., and Verhoef, H.A. (2004). Biodiversity effects on soil processes explained by interspecific functional dissimilarity. *Science* 306, 1019–1020.

Helgason, T., Daniell, T.J., Husband, R., Fitter, A.H., and Young, J.P.W. (1998). Ploughing up the wood-wide web? *Nature* 394, 431.

Hendrix, P.F., Parmelee, R.W., Crossley, D.A., Coleman, D.C., Odum, E.P., and Groffman, P.M. (1986). Detritus food webs in conventional and no-tillage agroecosystems. *BioScience* 36, 374–380.

Henry, H.A.L. and Jefferies, R.L. (2003). Plant amino acid uptake, soluble N turnover and microbial N capture in soils of a grazed Arctic salt marsh. *Journal of Ecology* 91, 627–636.

Herman, R.P., Provencio, K.R., Herrera-Matos, J., and Torrez, R.J. (1995). Resource islands predict the distribution of heterotrophic bacteria in Chihuahuan Desert soils. *Applied and Environmental Microbiology* 61, 1816–1821.

Hobbie, S.E. (1996). Temperature and plant species control over litter decomposition in Alaskan tundra. *Ecological Monographs* 66, 503–522.

Hobbie, S.E., Shevtsova, A., and Chapin, F.S. (1999). Plant responses to species removal and experimental warming in Alaskan tussock tundra. *Oikos* 84, 417–434.

Hodge, A., Stewart, J., Robinson, D., Griffiths, B.S., and Fitter, A.H. (1998). Root proliferation, soil fauna and plant nitrogen capture from nutrient-rich patches in soil. *New Phytologist* 139, 479–494.

Hodge, A. Stewart, J., Robinson, D., Griffiths, B.S., and Fitter, A.H. (1999). Plant, soil fauna and microbial responses to N-rich organic patches of contrasting temporal availability. *Soil Biology and Biochemistry* 31, 1517–1530.

Hodge, A. Stewart, J., Robinson, D., Griffiths, B.S., and Fitter, A.H. (2000). Competition between roots and soil microorganisms for nutrients from nutrient-rich patches of varying complexity. *Journal of Ecology* 88, 150–164.

Hodge, A., Campbell, C.D., and Fitter, A.H. (2001). An arbuscular mycorrhizal fungus accelerates decomposition and acquires nitrogen directly from organic material. *Nature* 413, 297–299.

Hodkinson, I.D., Coulson, S.J., Webb, N.R., and Block, W. (1996). Can high Arctic soil microarthropods survive elevated summer temperatures? *Functional Ecology*, 10, 314–321.

Hodkinson, I.D., Coulson, S.J., Harrison, J., and Webb, N.R. (2001). What a wonderful web they weave: spiders, nutrient capture and early ecosystem development in the high Arctic—some counter-intuitive ideas on community assembly. *Oikos* 95, 349–352.

Hoeksema, J.D., Lussenhop, J., and Teeri, J.A. (2000). Soil nematodes indicate food web responses to elevated atmospheric CO_2. *Pedobiologia* 44, 725–735.

Hokka, V., Mikola, J., Vestberg, M., and Setälä, H. (2004). Interactive effects of defoliation and an AM fungus on plants and soil organisms in experimental legume-grass communities. *Oikos* 106, 73–84.

Holland, E.A. and Detling, J.K. (1990). Plant response to herbivory and belowground nitrogen cycling. *Ecology* 71, 1040–1049.

Holland, J.N., Cheng, W., and Crossley, D.A. Jr. (1996.) Herbivore-induced changes in plant carbon allocation: assessment of below-ground C fluxes using carbon-14. *Oecologia* 107, 87–94.

Honer-Devine, M.C., Carney, K.M., and Bohannan, B.J.M. (2004). An ecological perspective on bacterial diversity. *Proceedings of the Royal Society London* (B) 271, 113–122.

Hoogerkamp, M., Rogaar, H., and Eijsackers, H.J.P. (1983). Effect of earthworms on grassland ion recently reclaimed polder soils in the Netherlands. In: *Earthworm Ecology: From Darwin to Vermiculture* (ed. J.E. Satchell), Chapman and Hall, London, pp. 85–105.

Hooper, U.H. and Vitousek, P.M. (1997). The effects of plant composition and diversity on ecosystem processes. *Science* 277, 1302–1305.

Hopp, H. and Slater, C.S. (1948). Influence of earthworms on soil productivity. *Soil Science* 66, 421–428.

Horton, T.R. and Bruns, T.D. (2001). The molecular revolution in ectomycorrhizal ecology: peeking into the black-box. *Molecular Ecology* 10, 1855–1871.

Houghton, J.T., Meira Filho, L.G., Callander, B.A., Harris, N., Kattenberg, A., and Maskell, K. (1995). *Climate Change 1995: The Science of Climate Change.* Cambridge University Press, Cambridge, UK

Hu, S., Chapin, F.S., Firestone, M.K., Field, C.B., and Chiariello, N.R. (2001). Nitrogen limitation of microbial decomposition in a grassland under elevated CO_2. *Nature* 409, 188–191.

Hudson, J.J., Dillon, P.J., and Somers, K.M. (2003). Long-term patterns in dissolved organic carbon in boreal lakes: the role of incident radiation, precipitation, air temperature, southern oscillation and acid deposition. *Hydrology and Earth System Science* 7, 390–398.

Hungate, B.A., Canadell, J., and Chapin, F.S. (1996). Plant species mediate changes in soil microbial N in response to elevated CO_2. *Ecology* 77, 2505–2515.

Hungate, B.A., Chapin, F.S., Zhong, H., Holland, E.A., and Field, C.B. (1997*a*). Stimulation of grassland nitrogen cycling under carbon dioxide enrichment. *Oecologia* 109, 149–153.

Hungate, B.A., Holland, E.A., Jackson, R.B., Chapin, F.S., Mooney, H.A., and Field, C.B. (1997*b*). The fate of carbon in grasslands under carbon dioxide enrichment. *Nature* 388, 576–579.

Hungate, B.A., Dijkstra, P., Johnson, D.W., Hinkle, C.R., and Drake, B.G. (1999). Elevated CO_2 increases nitrogen fixation and decreases soil nitrogen mineralization in Florida scrub oak. *Global Change Biology* 5, 781–789.

Hungate, B.A., Dukes, J.S., Shaw, M.R., Lou, Y., and Field, C.B. (2003). Nitrogen and climate change. *Science* 302, 1512–1513.

Hunt, H.W. and Wall, D.H. (2002). Modelling the effects of loss of soil biodiversity on ecosystem function. *Global Change Biology* 8, 33–50.

Husband, R., Herre, E.A., Turner, S.L., Gallery, R., and Young, J.P.Y. (2002). Molecular diversity of arbuscular mycorrhizal fungi and patterns of host association over time and space in a Tropical forest. *Molecular Ecology* 11, 2669–2678.

Huston, M.A. (1979). A general model of species diversity. *American Naturalist* 113, 81–101.

Huston, M.A. (1994). *Biological Diversity*. Cambridge University Press, Cambridge.

Ingham, R.E. and Detling, J.K. (1984). Plant–herbivore interactions in a North American mixed-grass prairie III. Soil nematode populations and root biomass on *Cynomys ludovicianus* colonies and adjacent uncolonized areas. *Oecologia* 63, 307–313.

Ingham, R.E. and Detling, J.K. (1990). Effects of root-feeding nematodes on aboveground net primary production in a North American grassland. *Plant and Soil* 121, 279–281.

Ingham, R.E., Trofymow, J.A., Ingham, E.R., and Coleman, D.C. (1985). Interactions of bacteria, fungi, and their nematode grazers: effects on nutrient cycling and plant growth. *Ecological Monographs* 55, 119–140.

Innes, L., Hobbs, P.J., and Bardgett, R.D. (2004). The impacts of individual plant species on rhizosphere microbial communities varies in soils of different fertility. *Biology and Fertility of Soils* 40, 7–13.

Insam, H. and Haselwandter, K. (1989). Metabolic quotient of the soil microflora in relation to plant succession. *Oecologia* 79, 174–178.

Intergovernmental Panel on Climate Change (2001). *Climate Change 2001: The Scientific Basis*. Cambridge University Press, NY.

Jackson, L.E., Schimel, D.S., and Firestone, M.K. (1989). Short-term partitioning of ammonium and nitrate between plants and microbes in an annual grassland. *Soil Biology and Biochemistry* 21, 409–415.

Jaeger, C.H., Monson, R.K., Fisk, M.C., and Schmidt, S.K. (1999). Seasonal partitioning of nitrogen by plants and soil microorganisms in an alpine ecosystem. *Ecology* 80, 1883–1891.

James, S.W. (1995). Systematics, biogeography, and ecology or nearctic earthworms from eastern, central, southern and southwestern United States. In: *Earthworm Ecology and Biogeography in North America* (ed. P.F. Hendrix), Boca Raton, Lewis, FL, pp. 29–52.

James, S.W. and Hendrix, P.F. (2004). Invasion of exotic earthworms into North America and other regions. In: *Earthworm Ecology*, (ed. C.A. Edwards), 2nd edn, CRC Press, Boca Raton, FL.

Jano, A., Jefferies, R.L., and Rockwell, R. (1998). The detection of vegetational change by multitemporal analysis of LANDSAT data: the effects of goose foraging. *Journal of Ecology* 86, 93–99.

Janssens, I.A., Crookshanks, M., Taylor, G., and Ceulemans, R. (1998). Elevated atmospheric CO_2 increases fine root production, respiration, rhizosphere respiration and soil CO_2 efflux in Scots pine seedlings. *Global Change Biology* 4, 871–878.

Jefferies, R.L. (1988). Pattern and process in arctic coastal vegetation in response to foraging by lesser snow geese. In: M. Werger, P. Van der Aart, H.J. During, and J. Verhoeven *Plant Form and Vegetation Structure, Adaptation, Plasticity and Relationship to Herbivory*. SPB Academic Publishers, The Hague (eds. pp. 281–300).

Jenkinson, D.S., Adams, D.E., and Wild, A. (1991). Model estimates of CO_2 emissions from soil in response to global warming. *Nature* 351, 304–306.

Jenny, H. (1941). *Factors of Soil Formation*. McGraw Hill, NY.

Jentschke, G., Bonkowski, M., Godbold, D.L., and Scheu. (1995). Soil protozoa and plant growth: non-nutritional effects and interaction with mycorrhizas. *Biology and Fertility of Soils* 20, 263–269.

Johnson, N.C., Graham, J.H., and Smith, F.A. (1997). Functioning of mycorrhizal associations along the mutualism–parasitism continuum. *New Phytologist* 135, 575–586.

Johnson, D., Leake, J.R., and Lee, J.A. (1998). Changes in soil microbial biomass and microbial activity in response to 7 years simulated pollutant nitrogen deposition on heathland and two grasslands. *Environmental Pollution* 103, 239–250.

Johnson, D.W., Cheng, W., and Ball, J.T. (2000). Effects of [CO_2] and nitrogen fertilization on soils planted with ponderosa pine. *Plant and Soil* 224, 99–113.

Johnson, D., Vandenkoornhuyse, P.J., Leake, J.R., Gilbert, L., Booth, R.E., Grime, J.P., et al. (2003*a*). Plant communities affect arbuscular mycorrhizal fungal diversity and community composition in grassland microcosms. *New Phytologist* 161, 503–515.

Johnson, D., Booth, R.E., Whiteley, A.S., Bailey, M.J., Read, D.J., Grime, J.P., et al. (2003*b*). Plant community composition affects the biomass, activity and diversity of microorganisms in limestone grassland soil. *European Journal of Soil Science* 54, 671–678.

Johnson, N.C., Rowland, D.L., Corkidi, L., Egerton-Warburton, L.M, and Allen, E.B. (2003*c*). Nitrogen enrichment alters mycorrhizal allocation at five mesic to semi-natural grasslands. *Ecology* 84, 1895–1908.

Jonasson, S., Michelsen, A., and Schmidt, I.K. (1999*a*). Coupling of nutrient cycling and carbon dynamics in the Arctic, integration of soil microbial and plant processes. *Applied Soil Ecology* 11, 135–146.

Jonasson, S., Michelsen, A., Schmidt, I.K., and Nielsen, E.V. (1999*b*). Responses in microbes and plants to changed temperature, nutrient, and light regimes in the Arctic. *Ecology* 80, 1828–1843.

Jonasson, S., Michelsen, A., Schmidt, I.K., and Nielsen, E.V. (1996). Microbial biomass C, N and P in two arctic soils and responses to addition of NPK fertilizer and sugar: implications for plant nutrient uptake. *Oecologia* 106, 507–515.

Jones, D.L. and Kielland, K. (2002). Soil amino acid turnover dominates the nitrogen flux in permafrost-dominated taiga forest soils. *Soil Biology and Biochemistry* 34, 209–219.

Jones, D.T., F.-X., S., Bignell, D.E., Suryo, H., Gillison, A.W., and Eggleton, P. (2003). Termite assemblage collapses along a land-use intensification gradient in lowland central Sumatra, Indonesia. *Journal of Applied Ecology* 40, 380–391.

Jones, M.B. and Donnelly, A. (2004). Carbon sequestration in temperate grass land ecosystems and the influence of management, climate and elevated CO_2. *New Phytologist* 164, 423–439.

Jones, T.H., Thompson, L.J., Lawton, J.H., Bezemer, T.M., Bardgett, R.D., Blackburn, T.M., et al. (1998). Impacts of rising atmospheric carbon dioxide on model terrestrial ecosystems. *Science* 280, 441–443.

Jonsson, L., Nilsson, M.-C., Wardle, D.A., and Zackrisson, O. (2001). Context dependent effects of ectomycorrhizal species richness on tree seedling productivity. *Oikos* 93, 353–364.

Kaiser, J. (2004). Wounding earth's fragile skin. *Science* 304, 1616–1618.

Kampichler, C., Kandeler, E., Bardgett, R.D., Jones, T.H., and Thompson, L.J. (1998). Impact of elevated atmospheric CO_2 concentration on soil microbial biomass and activity in a complex, weedy field model ecosystem. *Global Change Biology* 4, 335–346.

Kandeler, E., Tscherko, D., Bardgett, R.D., Hobbs, P.J., Kampichler, C., and Jones, T.H. (1998). The response of soil microorganisms and roots to elevated CO_2 and temperature in a terrestrial model ecosystem. *Plant and Soil* 202, 251–262.

Kane, E.S., Pregitzer, K.S., and Burton, A.J. (2003). Soil respiration along environmental gradients in Olympic National Park. *Ecosystems* 6, 326–335.

Kaufmann, R. (2001). Invertebrate succession on an alpine glacier foreland. *Ecology* 82, 2261–2278.

Kaye, J.P. and Hart, S.C. (1997). Competition for nitrogen between plants and soil microorganisms. *Trends in Ecology and Evolution* 12, 139–143.

Kennedy, I.R. (1992). *Acid Soil and Acid Rain*, 2nd edn., John Wiley, NY.

Kielland, K. (1994). Amino acid absorption by arctic plants: implication for plant nutrition and nitrogen cycling. *Ecology* 75, 2373–2383.

Kielland, K., Bryant, J.P., and Ruess, R.W. (1997). Moose herbivory and carbon turnover of early successional stands in interior Alaska. *Oikos* 80, 25–30.

King, L.K. and Hutchinson, K.J. (1976). The effects of sheep stocking intensity on the abundance and distribution of mesofauna in pastures. *Journal of Applied Ecology* 13, 41–55.

King, L.K., Hutchinson, K.J., and Greenslade, P. (1976). The effects of sheep numbers on associations of Collembola in sown pastures. *Journal of Applied Ecology* 13, 731–739.

King, R., Dromph, K., and Bardgett, R.D. (2002). Changes in species evenness of litter have no effect on decomposition processes. *Soil Biology and Biochemistry* 34, 1959–1963.

Klamer, M., Roberts, M.S., Levine, L.H., Drake, B.G., and Garland, J.L. (2002). Influence of elevated CO_2 on the fungal community in a coastal scrub oak forest soil investigated with terminal-restriction fragment length polymorphism analysis. *Applied and Environmental Microbiology* 68, 4370–4376.

Klironomos, J.N. (2002). Feedback to the soil community contributes to plant rarity and invasiveness in communities. *Nature* 417, 67–70.

Klironomos, J.N. (2003). Variation in plant response to native and exotic arbuscular mycorrhizal fungi. *Ecology* 84, 2292–2301.

Klironomos, J.N., Rillig, M.C., Allen, M.F., Zak, D.R., Kubiske, M., and Pregitzer, K.S. (1997). Soil fungal-arthropod responses to *Populus tremuloides* grown under enriched atmospheric CO_2 under field conditions. *Global Change Biology* 3, 473–478.

Knight, D., Elliot, P.W., Anderson, J.M., and Scholfield, P. (1992). The role of earthworms in managed, permanent pastures in Devon, England. *Soil Biology and Biochemistry* 24, 1511–1517.

Kohls, S.J., Baker, D.D., van Kessel, C., and Dawson, J.O. (2003). An assessment of soil enrichment by actinorhizal N_2 fixation using delta ^{15}N values in a chronosequence of deglaciation at Glacier Bay, Alaska. *Plant and Soil* 254, 11–17.

Körner, C. (1999). *Alpine Plant Life*. Springer-Verlag, Berlin.

Körner, C. and Arnone, J.A. (1992). Responses to elevated carbon-dioxide in artificial tropical ecosystems. *Science* 257, 1672–1675.

Körner, C., Diemer, M., Schäppi, B., Niklause, P.A., and Arnone, J. (1997). The response of alpine grassland to four seasons of CO_2 enrichment: a synthesis. *Acta Oecologia* 18, 165–175.

Kowalchuck, G.A., Stephen, J.R., de Boer, W., Prosser, J.I., Embley, T.M., and Woldendorp, J.W. (1997). Analysis of ammonia-oxidising bacteria of the β-subdivision of the class *Proteobacteria* in coastal sand dunes by denaturing gradient gel electrophoresis and sequencing of PCR-amplified 16S ribosomal DNA fragments. *Applied and Environmental Microbiology* 63, 1489–1497.

Krebs, K.J. (2001) *Ecology*, 5th Edition. Benjamin Cummings, San Francisco.

Kuikman, P.J., Jansen, A.G., van Veen, J.A., and Zehnder, A.J.B. (1990). Protozoan predation and the turnover of soil organic carbon and nitrogen in the presence of plants. *Biology and Fertility of Soils* 10, 22–28.

Laakso, J. and Setälä, H. (1999*a*). Population- and ecosystem-level effects of predation on microbial-feeding nematodes. *Oecologia* 120, 279–286.

Laakso, J. and Setälä, H. (1999*b*). Sensitivity of primary production to changes in the architecture of belowground food webs. *Oikos* 87, 57–64.

Lachenbruch, A.H. and Marshall, B.V. (1986). Changing climate: geothermal evidence from permafrost in the Alaskan Arctic. *Science* 234, 689–696.

Lal, R. (1997). Long-term tillage and maize monoculture effects on a tropical Alfisol in Western Nigeria. II. Soil chemical properties. *Soil and Tillage Research* 42, 161–174.

Lal, R. (2002). *Encyclopaedia of Soil Science*. Marcel Dekker, NY, USA.

Lal, R. (2004). Soil carbon sequestration impacts on global climate change and food security. *Science* 304, 1623–1627.

Lange, O.L. (2001). Photosynthesis of soil-crust biota as dependant on environmental factors. In: *Biological Soil Crusts: Structure, Function and Management*, (eds. J. Belnap and O.L. Lange), Springer-Verlag, Berlin, pp. 217–240.

Lapied, E. and Lavelle, P. (2003). The peregrine earthworm *Pontoscolex corethrurus* populations in the East Coast of Costa Rica. *Pedobiologia* 47, 471–474.

Larsen, K.S., Jonasson, S., and Michelsen, A. (2002). Repeated freeze–thaw cycles and their effects on biological processes in two arctic ecosystem types *Applied Soil Ecology* 21, 187–195.

Larsen, N., Olsen, G.J., Maidak, B.L., McCaughey, M.J., Overbeek, R., Macke, T.J., et al. (1993). The ribosomal database project. *Nucleic Acid Research* 21, 3021–3023.

Larson, J.L. (2000). Elevated CO_2, O_3 and the response of microbial communities. *School of Natural Resources and Environment*. Ann Arbor, MI.

Lavelle, P. and Martin, A. (1992). Small-scale and large-scale effects of endogeneic earthworms on soil organic matter dynamics in soil and the humid tropics. *Soil Biology and Biochemistry* 24, 1491–1498.

Lavelle, P. and Pashanasi, B. (1989). Soil macrofauna and land management in Peruvian Amazonia (Yurimaguas, Loreto). *Pedobiologia* 33, 283–291.

Lavelle, P., Lattaud, C., Trigo, D., and Barois, I. (1995). Mutualism and biodiversity in soils. *Plant and Soil* 170, 23–33.

Lavelle, P., Bignell, D., Lepage, M., Wolters, V., Roger, P., Ineson, P., et al. (1997). Soil function in a changing world: the role of invertebrate ecosystem engineers. *European Journal of Soil Biology* 33, 159–193.

Lawton, J.H., Bignell, D.E., Bloemers, G.F., Eggleton, P., and Hodda, M.E. (1996). Carbon flux and diversity of nematodes and termites in Cameroon forest soils. *Biodiversity and Conservation* 5, 261–273.

Leake, J.R. and Read, D.J. (1989). The effects of phenolic compounds on nitrogen mobilization by ericoid mycorrhizal systems. *Agriculture, Ecosystems, and Environment* 29, 225–236.

Leake, J.R. and Read, D.J. (1990). Proteinase activity in mycorrhizal fungi. I. The effect of extracellular pH on the production and activity of proteinase by ericoid endophytes from soils of contrasting pH. *New Phytologist* 115, 243–250.

Leake, J.R., Johnson, D., Donnelly, D.P., Boddy, L., and Read, D.J. (2005). Is diversity of mycorrhizal fungi important for ecosystem functioning? In: *Biological Diversity and Function in Soil*, (eds. R.D. Bardgett, M.B. Usher, and D.W. Hopkins), Cambridge University Press, Cambridge, UK.

Lee, K. (1985). *Earthworms: Their Ecology and Relationships with Soils and Land Use*. Academic Press, NY, USA.

Leonard, M.A. (1984). Observations on the influence of culture conditions on the fungal preference of *Folsomia candida* (Collembola; Isotomidae). *Pedobiologia* 26, 361–367.

Liiri, M., Setälä, H., Haimi, J., Pennanen, T., and Fritze, H. (2001). Relationship between soil microarthropod species diversity and plant growth does not change when the system is disturbed. *Oikos* 96, 138–150.

Liiri, M., Setälä, H., Haimi J., Pennanen, T., and Fritze, H. (2002). Soil processes are not influenced by the functional complexity of soil decomposer food webs under disturbance. *Soil Biology and Biochemistry* 34, 1009–1020.

Lilleskov, E.A., Fahey, T.J., and Lovett, G.M. (2001). Ectomycorrhizal fungal aboveground community change over an atmospheric nitrogen deposition gradient. *Ecological Applications* 11, 397–410.

Lipson, D.A. and Monson, R.K. (1998). Plant-microbe competition for soil amino acids in the alpine tundra: effects of freeze–thaw and dry–rewet events. *Oecologia* 113, 406–414.

Lipson, D.A. and Näsholm, T. (2001). The unexpected versatility of plants: organic nitrogen use and availability in terrestrial ecosystems. *Oecologia* 128, 305–316.

Lloyd, J.(1999). The CO_2 dependence of photosynthesis, plant growth responses to elevated CO_2 concentrations and their interactions with soil nutrient status. II. Temperature and boreal forest productivity and the combined effects of increasing CO_2 concentrations and increased nitrogen deposition. *Functional Ecology* 13, 439–459.

Lobry De Bruyn, L.A. and Conacher, A.J. (1994). The effect of ant biopores on water infiltration in soils in undisturbed brushland and farmland in a semi-arid environment. *Peodobiolgia* 38, 193–207.

Loreau, M., Naeem, S., and Inchausti, P. (2002). *Biodiversity and Ecosystem Functioning: Synthesis and Perspectives*, Oxford University Press, Oxford.

Lovett, G.M. and Ruesink, A.E. (1995). Carbon and nitrogen mineralization from decomposing gypsey moth frass. *Oecologia* 104, 133–138.

Lundkvist, H. (1983). Effects of clear-cutting on the enchytraeids in a Scots pine forest soil in central Sweden. *Journal of Applied Ecology* 20, 873–885.

Luo, Y., Su, B., Currie, W.S., Dukes, J.S., Finzi, A., Hartwig, U., et al. (2004). Progressive nitrogen limitation responses to rising atmospheric carbon dioxide. *BioScience* 54, 731–739.

Lussenhop, J., Treonis, A., Curtis, P.S., Teeri, J.A., and Vogel, C.S. (1998). Response of soil biota to elevated atmospheric CO_2 in poplar model systems. *Oecologia* 113, 247–251.

MacDonald, W.H., Dise, N.B., Matzner, E., Armbruster, M., Gunderson, P., and Forsius, M. (2002). Nitrogen input together with ecosystem nitrogen enrichment predict nitrate leaching from European forests. *Global Change Biology* 8, 1028–1033.

Mack, M.C., Schuur, E.A.G., Bret-Harte, M.S., Shaver, G.R., and Chapin, F.S. (2004). Ecosystem carbon storage in arctic tundra reduced by long-term nutrient fertilization. *Nature* 431, 440–443.

Manseau, M., Huot, J., and Crête, M. (1996). Effects of summer grazing by caribou on composition and productivity of vegetation: community and landscape level. *Journal of Ecology* 84, 503–513.

Maraun, M., Martens, H., Migge, S., Theenhaus, A., and Scheu, S. (2003). Adding to 'the enigma of soil animal diversity: fungal feeders and saprophagous soil invertebrates prefer similar food substrates. *European Journal of Soil Biology* 39, 85–95.

Markkola, A.M., Ohtonen, A., AhonenJonnarth, U., and Ohtonen, R. (1996). Scots pine responses to CO_2 enrichment-1. Ectomycorrhizal fungi and soil fauna. *Environmental Pollution* 94, 309–316.

Martin-Olmedo, P., Rees, R.M., and Grace, J. (2002). The influence of plants grown under elevated CO_2 and N fertilization on soil nitrogen dynamics. *Global Change Biology* 8, 643–657.

Masters, G.J. and Brown, V.K. (1992). Plant-mediated interactions between two spatially separated insects. *Functional Ecology* 6, 175–179.

Masters, G.J., Brown, V.K., and Gange, A.C. (1993). Plant mediated interactions between above- and below-ground insect herbivores. *Oikos* 66, 148–151.

Masters, G.J., Jones, T.H., and Rogers, M. (2001). Host-plant mediated effects of root herbivory on insect seed predators and their parasitoids. *Oecologia* 127, 246–250.

Matthews, J.A. (1992). *The Ecology of Recently Deglaciated Terrain*, Cambridge University Press, Cambridge.

Mawdsley, J.L. and Bardgett, R.D. (1997). Continuous defoliation of perennial ryegrass (*Lolium perenne*) and white clover (*Trifolium repens*) and associated changes in the microbial population of an upland grassland soil. *Biology and Fertility of Soils* 24, 52–58.

Maxwell, B. (1992). Arctic climate: a potential for change under global warming. In: *Arctic Ecosystems in a Changing Climate: An Ecophysiological Perspective* (eds. F.S. Chapin, R.L. Jefferies, J.F. Reynolds, G.R. Shaver, and J. Svoboda), Academic Press, San Diego, CA, pp. 11–34.

McArthur, R.H. and Wilson, E.O. (1967). *The Theory of Island Biogeography*, Princeton University Press, Princeton, NJ US.

McClaugherty, C.A. (1983). Soluble polyphenols and carbohydrates in throughfall and leaf litter decomposition. *Acta Oecologia* 4, 375–385.

McDowell, W.H., Currie, W.S., Aber, J.D., and Yano, Y. (1998). Effects of chronic nitrogen amendments on production of dissolved organic carbon and nitrogen in forest soils. *Water Air and Soil Pollution* 105, 175–182.

McKane, R.B., Rastetter, E.B., Shaver, G.R., Nadelhoffer, K.J., Giblin, A.E., Laundre, J.A., et al. (1997a). Climatic effects on tundra carbon storage inferred from experimental data and a model. *Ecology* 78, 1170–1187.

McKane, R.B., Tingley, D.T., and Beedlow, P.A. (1997b). Spatial and temporal scaling of CO_2 and temperature effects on Pacific Northwest forest ecosystems. *American Association for Advances is Science, Pacific Division Abstracts* 16, 56.

McKane, R.B., Johnson, L.C., Shaver, G.R., Nadelhoffer, K.J., Rastetterk, E.B., Fry B., et al. (2002). Resource-based niches provide a basis for plant species diversity and dominance in arctic tundra. *Nature* 413, 68–71.

McNaughton, S.J. (1985). Ecology of a grazing system: the Serengeti. *Ecological Monographs* 55, 259–294.

McNaughton, S.J., Osterheld, M., Frank, D.A., and Williams, K.J. (1989). Ecosystem level patterns of primary productivity and herbivory in terrestrial habitats. *Nature* 341, 142–144.

McNaughton, S.J., Banyikwa, F.F., and McNaughton, M.M. (1997a). Promotion of the cycling of diet-enhancing nutrients by African grazers. *Science* 278, 1798–1800.

McNaughton, S.J., Zuniga, G., McNaughton, M.M., and Banyikwa, F.F., (1997b). Ecosystem catalysis: soil urease activity and grazing in the Serengeti ecosystem. *Oikos* 80, 467–469.

McNaughton, S.J., Banyikwa, F.F., and McNaughton, M.M. (1998). Root biomass and productivity in a grazing ecosystem: the Serengeti. *Ecology* 79, 587–592.

McNeill, J.R. and Winiwarter, V. (2004). Breaking the sod: mankind, history, and soil. *Science* 304, 1627–1629.

McRill, M. (1974). The ingestion of weed seeds by earthworms. In: *The Proceedings of the 12th British Weed Control Conference*, British Crop Protection Council, London, pp. 519–524.

Mikan, C.J., Zak, D.R., Kubiske, M.E., and Pregitzer, K.S. (2000). Combined effects of atmospheric CO_2 and N availability on the belowground carbon and nitrogen dynamics of aspen mesocosms. *Oecologia* 124, 432–445.

Mikola, J. and Setälä, H. (1998a). Productivity and trophic-level biomasses in a microbial-based soil food web. *Oikos* 82, 158–168.

Mikola, J. and Setälä, H. (1998b). No evidence of trophic cascades in an experimental microbial-based soil food web. *Ecology* 79, 153–164.

Mikola, J., Yeates, G.W., Barker, G.M., Wardle, D.A., and Bonner, K.I. (2001). Effects of defoliation intensity on soil food-web properties in an experimental grassland community. *Oikos* 92, 333–343.

Mikola, J., Bardgett, R.D., and Hedlund, K. (2002). Biodiversity, ecosystem functioning and soil decomposer food webs. In: *Biodiversity and Ecosystem Functioning: Synthesis and Perspectives* (eds. M. Loreau, S. Naeem, and P. Inchausti), Oxford University Press, pp. 169–180.

Milchunas, D.G. and Lauenroth, W.K. (1993). Quantitative effects of grazing on vegetation and soils over a global range of environments. *Ecological Monographs* 63, 327–366.

Miller, A.E. and Bowman, W.D. (2002). Variation in nitrogen-15 natural abundance and nitrogen uptake traits among co-occurring alpine species: do species partition by nitrogen form? *Oecologia* 130, 609–616.

Miller, A.E. and Bowman, W.D. (2003). Alpine plants show species-level differences in the uptake of organic and inorganic nitrogen. *Plant and Soil* 250, 283–292.

Mills, K.E. and Bever, J.D. (1998). Maintenance of diversity within plant communities: soil pathogens as agents of negative feedback. *Ecology* 79, 1595–1601.

Mittelbach, G.G., Steiner, C.F., Scheiner, S.M., Gross, K.L., Reynolds, H.L., Waide, R.B., et al. (2001). What is the observed relationship between species richness and productivity? *Ecology* 82, 2381–2396.

Molvar, E.M., Bowyer, R.T., and van Ballenberghe, V. (1993). Moose herbivory, browse quality, and nutrient cycling in an Alaskan treeline community. *Oecologia* 94, 472–479.

Montealegre, C.M., van Kessel, C., Russelle, M.P., and Sadowsky, M.J. (2002). Changes in microbial activity and composition in a pasture ecosystem exposed to elevated atmospheric carbon dioxide. *Plant and Soil* 243, 197–207.

Moody, S.A., Piearce, T.G., and Dighton, J. (1996). Fate of some fungal spores associated with wheat straw decomposition on passage through the guts of *Lumbricus terrestris* and *Aporrectodea longa*. *Soil Biology and Biochemistry* 28, 533–537.

Moore, J.C. and Hunt, H.W. (1988). Resource compartmentation and the stability of real ecosystems. *Nature* 333, 261–263.

Moore, J.C. and de Ruiter, P.C. (2000). Invertebrates in detrital food webs along gradients of productivity. In: *Invertebrates as Webmasters in Ecosystems* (eds. D.C. Coleman and P.F. Hendrix), CABI Publishing, Oxford, NY, pp. 161–184.

Moore, J.C., St. John, T.V., and Coleman, D.C. (1985). Ingestion of vesicular-arbuscular mycorrhizal hyphae and spores by soil microarthropods. *Ecology* 66, 1979–1981.

Moran, N.A. and Whitham, T.G. (1990). Interspecific competition between root-feeding and leaf-galling aphids mediated by host-plant resistance. *Ecology* 71, 1050–1058.

Mountford, J.O., Lakhani, K.H., and Kirkham, F.W. (1993). Experimental assessment of the effects of nitrogen addition under hay-cutting and aftermath grazing on the vegetation of meadows on a Somerset peat moor. *Journal of Applied Ecology* 30, 321–332.

Murray, P.J., Ostle, N., Kenny, C., and Grant, H. (2004). Effect of defoliation on patterns of carbon exudation from *Agrostis capillaris*. *Journal of Plant Nutrition and Soil Science* 167, 487–493.

Nadelhoffer, K.L. and Fry, B. (1994). Controls on nitrogen-15 and carbon-13 abundances in forest soil organic matter. *Soil Science Society of America Journal* 52, 1633–1640.

Nadelhoffer, K.L., Aber, J.D., and Melillo, J.M. (1985). Fine roots, net primary production and soil nitrogen availability. *Ecology* 66, 1377–1390.

Nadelhoffer, K.L., Giblin, A.E., Shaver, G.R., and Launder, A.E., (1991). Effects of temperature and substrate quality on element mineralization in six arctic soils. *Ecology* 72, 242–253.

Näsholm, T., Ekblad, A., Nordin, A., Giesler, R., Högberg, M., and Högberg, P. (1998). Boreal forest plants take up organic nitrogen. *Nature* 392, 914–916.

Näsholm, T., Huss-Danell, K., and Högberg, P. (2000). Uptake of organic nitrogen in the field by four agriculturally important plant species. *Ecology* 81, 1155–1161.

Näsholm, T., Huss-Danell, K., and Högberg, P. (2001). Uptake of glycine by field grown wheat. *New Phytologist* 150, 59–63.

Neff, J.C., Townsend, A.R., Gleixner, G., Lehmann, S.J., Turnbull, J., and Bowman, W.D. (2002). Variable effects of nitrogen additions on the stability and turnover of soil carbon. *Nature* 419, 915–917.

Newbould, P. (1982). Biological nitrogen fixation in upland and marginal areas. *Philosophical Transactions of the Royal Society, London B* 296, 405–417.

Newell, K. (1984*a*). Interaction between two decomposer basidiomycetes and a collembolan under Sitka spruce: grazing and its potential effects on fungal distribution and litter decomposition. *Soil Biology and Biochemistry* 16, 235–239.

Newell, K. (1984*b*). Interaction between two decomposer basidiomycetes and a collembolan under Sitka spruce: distribution, abundance and selective grazing. *Soil Biology and Biochemistry* 16, 227–233.

Newington, J.E., Setälä, H., Bezemer, T.M., and Jones, T.H. (2004). Potential effects of earthworms on leaf-chewer performance. *Functional Ecology* 18, 746–751.

Newsham, K.K., Fitter, A.H., and Watkinson, A.R. (1994). Root pathogenic and arbuscular mycorrhizal fungi determine fecundity of asymptomatic plants in the field. *Journal of Ecology* 82, 805–814.

Newsham, K.K., Fitter, A.H., and Watkinson, A.R. (1995). Multi-functionality and biodiversity in arbuscular mycorrhizas. *Trends in Ecology and Evolution* 10, 407–411.

Newton, P.C.D., Clark, H., Bell, C.C., Glasgow, E.M., Tate, K.R., Ross, D.J., et al. (1995). Plant growth and soil processes in temperate grassland communities at elevated CO_2. *Journal of Biogeography* 22, 235–240.

Nielson, R.L. and Hole, F.E. (1964). Earthworms and the development of coprogenous A1 horizons in forest soils of Wisconsin. *Soil Science Society of America, Proceedings* 28, 426–430.

Niklaus, P.A. and Korner, C. (1996). Responses of soil microbiota of a late successional alpine grassland to long term CO_2 enrichment. *Plant and Soil* 184, 219–229.

Niklaus, P.A., Kandeler, E., Leadley, P.W., Schmid, B., Tscherko, D., and Korner, C. (2001). A link between plant diversity, elevated CO_2 and soil nitrate. *Oecologia* 127, 540–548.

Nikolyuk, V.F. and Tapilskaja, N.V. (1969). Bodenamöben als Produzenten von biotisch aktiven Stoffen. *Pedobiologia* 9, 182–187.

Nohrstedt, H.Ö. and Börjesson G. (1998). Respiration in a forest soil 27 years after fertilization with different doses of urea. *Silva Fennica* 32, 383–388.

Nordin, A., Högberg, P., and Näsholm, T. (2001). Soil nitrogen form and plant nitrogen uptake along a boreal forest productivity gradient. *Oecologia* 129, 125–132.

Northup, R.R., Yu, Z., Dahlgren, R.A., and Vogt, K.A. (1995). Ployphenol control on nitrogen release from pine litter. *Nature* 377, 227–229.

Nunan, N., Ritz, K., Crabb, D., Harris, K., Wu, K., Crawford, J.W., et al. (2001). Quantification of the in situ distribution of soil bacteria by large-scale imaging of thin sections of undisturbed soil. *FEMS Microbiology Ecology* 37, 67–77.

O'Donnell, A.G., Colvan, S.R., Malosso, E., and Supaphol, S. (2005). Twenty years of molecular analysis of bacterial communities in soils and what have we learned about function? In: *Biological Diversity and Function in Soil* (eds. R.D. Bardgett, M.B. Usher, and D.W. Hopkins), Cambridge University Press, Cambridge, UK.

Oechel, A.C., Cook, A.C., Hastings, S.J., and Vourlitis, G.L. (1997). Effects of CO_2 and climate change on arctic ecosystems. In: *Ecology and Arctic Environments* (eds. S.J. Woodin and M. Marquiss), Special Publication Number 13 of the British Ecological Society, Blackwell Science, Oxford, pp. 255–273.

Oechel, W.C., Hastings, S.J., Vourlitis, G.L., Jenkins, M., Riechers, G., and Grulke, N. (1993). Recent change of Arctic tundra ecosystems from a net carbon dioxide sink to a source. *Nature* 361, 520–523.

Ohtonen, R., Fritze, H., Pennanen, T., Jumpponen, A., and Trappe, J. (1999). Ecosystem properties and microbial community changes in primary succession on a glacier forefront. *Oecologia* 119, 239–246.

Olff, H., Huisman, J., and van Tooren, B.F. (1993). Species dynamics and nutrient accumulation during early primary succession in coastal sand dunes. *Journal of Ecology* 81, 693–706.

Olff, H., Hoorens, B., De Goede, R.G.M., Van der Putten, W.H., and Gleichman, J.M. (2000). Small-scale shifting mosaics of two dominant grassland species: the possible role of soil-borne pathogens. *Oecologia* 125, 45–54.

Olofsson, J. and Oksanen, L. (2002). Role of litter decomposition for the increased primary production in areas heavily grazed by reindeer: a litterbag experiment. *Oikos* 96, 507–512.

Olofsson, J., Stark, S., and Oksanen, L. (2004). Reindeer influence on ecosystem processes in the tundra. *Oikos* 105, 386–396.

Oren, R., Ellsworth, D.S., Johnsen, K.H., Phillips, N., Ewers, B.E., Maier, C., et al. (2001). Soil fertility limits carbon sequestration by forest ecosystems in a CO_2-enriched atmosphere. *Nature* 411, 469–472.

Osterheld, M., Sala, O.E., and McNaughton, S.J. (1992). Effect of animal husbandry on herbivore-carrying capacity at a regional scale. *Nature* 356, 234–236.

Ostle, N., Whiteley, A.S., Bailey, M.J., Sleep, D., Ineson, P., and Manefield, M. (2003). Active microbial RNA turnover in a grassland soil estimated using a (CO_2)-C-13 spike. *Soil Biology and Biochemistry* 35, 877–885.

Owen, A.G. and D.L. Jones. (2001). Competition for amino acids between wheat roots and rhizosphere microorganisms and the role of amino acids in plant N acquisition. *Soil Biology and Biochemistry* 33, 651–657.

Pace, N.A., Stahl, D.A., Lane, D.J., and Olson, G. (1986). The analysis of natural microbial populations by ribosomal RNA sequences. *Advances in Microbial Ecology* 21, 59–68.

Packer, A. and Clay, K. (2000). Soil pathogens and spatial patterns of seedling mortality in a temperate tree. *Nature* 404, 278–281.

Pan, Y., Melillo, J.M., McGuire, A.D., Kicklighter, D.W., Pitelka, L.F., Hibbard, K., et al. (1998). Modeled responses of terrestrial ecosystems to elevated atmospheric CO_2: a comparison of simulations by biogeochemistry models of the Vegetation/ Ecosystem Modeling and Analysis Project (VEMAP). *Oecologia* 114, 389–404.

Parker, L.W., Santos, P.F., Phillips, J., and Whitford, W.G. (1984). Carbon and nitrogen dynamics during decomposition of litter and roots of a Chihuahuan desert annual *Lepidium lasiocarpum*. *Ecological Monographs* 54, 339–360.

Parkinson, D., Visser, S., and Whittaker, J.B. (1979). Effects of Collembolan grazing on fungal colonization of leaf litter. *Soil Biology and Biochemistry* 11, 529–535.

Pastor, J., Naiman, R.J., Dewey, B., and McInnes P. (1988). Moose, microbes and the boreal forest. *BioScience* 38, 770–777.

Pastor, J., Dewey, R.J., Naiman, R.J., McInnes, R.J., and Cohen, Y. (1993). Moose browsing and soil fertility in the boreal forests of Isle Royale National Park. *Ecology* 74, 467–480.

Paterson, E. and Sim, A. (1999). Rhizodeposition and C-partitioning of *Lolium perenne* in axenic culture by nitrogen supply and defoliation. *Plant and Soil* 216, 155–164.

Paterson, E., Rattray, E.A.S., and Killham, K. (1996). Effect of elevated atmospheric CO_2 concentration on C- partitioning and rhizosphere C-flow for three plant species. *Soil Biology and Biochemistry* 28, 195–201.

Paul, E.A. and Clark, F.E. (1996). *Soil Microbiology and Biochemistry*. Academic Press, San Diego, CA, USA.

Pearce, I.S.K. and Van der Wal, R. (2002). Effects of nitrogen deposition on growth and survival of montane *Racomitrium lanuginosum* heath. *Biological Conservation* 104, 83–89.

Pearce, I.S.K., Woodin, S.J., and Van der Wal, R. (2003). Physiological and growth responses of the montane bryophyte *Racomitrium lanuginosum* to atmospheric nitrogen deposition. *New Phytologist* 160,145–155.

Perakis, S.S. and Hedin, L.O. (2002). Nitrogen loss from unpolluted South American forests mainly via dissolved organic compounds. *Nature* 415, 416–419.

Petersen, H. and Luxton, M. (1982). A comparative analysis of soil fauna populations and their role in decomposition processes. *Oikos* 39, 287–388.

Phillips, R.L., Zak, D.R., Holmes, W.E., and White, D.C. (2002). Microbial community composition and function beneath temperate trees exposed to elevated atmospheric carbon dioxide and ozone. *Oecologia* 131, 236–244.

Phoenix, G.K., Booth, R.E., Leake, J.R., Read, D.J., Grime, J.P., and Lee, J.A. (2004). Effects of enhanced nitrogen deposition and phosphorus limitation on nitrogen budgets of semi-natural grasslands. *Global Change Biology* 9, 1309–1321.

Pickard, J. (1984). Exotic plants on Lord Howe Island: distribution in time and space 1853–1981. *Journal of Biogeography* 11, 181–208.

Piearce, T.G., Roggero, N., and Tipping, R. (1994). Earthworms and seeds. *Journal of Biological Education* 28, 195–202.

Ponge, J.F. (2003). Humus forms in terrestrial ecosystems: a framework to biodiversity. *Soil Biology and Biochemistry* 35, 935–945.

Ponsard, S. and Arditi, R. (2000). What can stable isotopes ($\delta^{15}N$ and $\delta^{13}C$) tell about the food web of soil macro-invertebrates? *Ecology* 81, 852–864.

Porazinska, D.L., Bardgett, R.D., Blaauw, M.B., Hunt, W.H., Parsons, A., Seastedt, T.R., et al. (2003). Relationships at the aboveground–belowground interface: plants, soil microflora and microfauna, and soil processes. *Ecological Monographs* 73, 377–395.

Post, D.M. (2002). Using stable isotopes to estimate trophic position: models, methods, and assumptions. *Ecology* 83, 703–718.

Post, W.M., Emmanuel, W.R., Zinke, P.J., and Stangenberger, A.G. (1982). Soil carbon pools and world life zones. *Nature* 298, 156–159.

Pregitzer, K.S., Zak, D.R., Burton, A.J., Ashby, J.A., and MacDonaldn, N.W. (2004). Chronic nitrate additions dramatically increase the export of carbon and nitrogen from northern hardwood ecosystems. *Biogeochemistry* 68, 179–197.

Press, M.C. (1998). Dracula or Robin Hood? A functional role for root hemiparasites in nutrient poor ecosystems. *Oikos* 82, 609–611.

Press, M.C. and Graves, J.D. (1995). *Parasitic Plants*. Chapman and Hall, London.

Press, M.C., Potter, J.A., Burke, M.J.W., Callaghan, T.V., and Lee, J.A. (1998). Responses of a subarctic dwarf shrub heath community to simulated environmental change. *Journal of Ecology* 86, 315–327.

Pywell, R.F., Bullock, J.M., Walker, K.J., Coulson, S.J., Gregory, S.J., and Stevenson, M.J. (2004). Facilitating grassland diversification using the hemiparasitic plant *Rhinanthus minor*. *Journal of Applied Ecology* 41, 880–887.

Quested, H.M., Press, M.C., Callaghan, T.V., and Cornelissen, J.H.C. (2002). The hemiparasite angiosperm *Bartsia alpina* has the potential to accelerate decomposition in sub-arctic communities. *Oecologia* 130, 88–95.

Quested, H.M., Press, M.C., and Callaghan, T.V. (2003*a*). Litter of the hemiparasite *Bartsia alpina* enhances plant growth: evidence for a functional role in nutrient cycling. *Oecologia* 135, 606–614.

Quested, H.M., Cornelissen, J.H.C., Press, M.C., Callaghan, T.V., Aerts, R., Trosien, F., et al. (2003*b*). Decomposition of sub-arctic plants with differing nitrogen economies: a functional role for hemiparasites. *Ecology* 84, 3209–3221.

Raab, T.K., Lipson, D.A., and Monson, R.K. (1999). Soil amino acid utilization among species of the cyperaceae: plant and soil processes. *Ecology* 80, 2408–2419.

Radajewski, S., Ineson, P., Parekh, N.R., and Murrell, J.C. (2000). Stable-isotope probing as a tool in microbial ecology. *Nature* 403, 646–649.

Raich, J.W. and Schlesinger, W.H. (1992). The global carbon dioxide flux in soil respiration and its relationship to vegetation and climate. *Tellus* 44B, 81–99.

Ratledge, C. and Wilkinson, S.G. (1988). *Microbial Lipids*, Vol. 1. Academic Press, NY.

Read, D.J. (1983). The biology of mycorrhiza in the Ericales. *Canadian Journal of Botany* 61, 985–1004.

Read, D.J. (1994). Plant-microbe mutualisms and community structure. In: *Biodiversity and Ecosystem Function* (eds. E.D. Schulze and H.A. Mooney), Springer-Verlag, Berlin, pp. 181–209.

Read, D.J. (1996). Plant-microbe mutualisms and community structure. In: *Biodiversity and Ecosystem Function* (eds. E.D. Shulze, and H.A. Mooney), Springer-Verlag, Berlin. pp. 181–209.

Reich, P.B., Knops, J., Tilman, D., Craine, J., Ellsworth, D., Tjoelker, M., et al. (2001*a*). Plant diversity enhances ecosystem responses to elevated CO_2 and nitrogen deposition. *Nature* 410, 809–812.

Reich, P.B., Tilman, D., Craine, J., Ellsworth, D., Tjoelker, M.G., Knops, J., et al. (2001*b*). Do species and functional groups differ in acquisition and use of C, N, and water under varying atmospheric CO_2 and N availability regimes? A field test with 16 grassland species. *New Phytologist* 150, 435–448.

Reinhart, K.O., Packer, A., Van der Putten, W.H., and Clay, K. (2003). Plant-soil biota interactions and spatial distribution of black cherry in its native and invasive ranges. *Ecology Letters* 6, 1046–1050.

Reynolds, R., Belnap, J., and Reheis, M. (2001). Aeolian dust in Colorado Plateau soils: nutrient inputs and recent change in source. *Proceedings of the National Academy of Science* 98, 7123–7127.

Reynolds, H.L., Packer, A., Bever, J.D., and Clay, K. (2003). Grassroots ecology: plant-microbe-soil interactions as drivers of plant community structure and dynamics. *Ecology* 84, 2281–2291.

Rhoades, D.F. (1985). Offensive–defensive interactions between herbivores and plants: their relevance in herbivore population dynamics and ecological theory. *American Naturalist* 125, 205–238.

Rice, C.W., Garcia, F.O., Hampton, C.O., and Owensby, C.E. (1994). Soil microbial response in tallgrass prairie to elevated CO_2. *Plant and Soil* 165, 67–74.

Rillig, M.C., Field, C.B., and Allen, M.F. (1999). Soil biota responses to long-term atmospheric CO_2 enrichment in two California annual grasslands. *Oecologia* 119, 572–577.

Ritchie, M.E., Tilman, D., and Knops, J.M.H. (1998). Herbivore effects on plant and nitrogen dynamics in oak savanna. *Ecology* 79, 165–177.

Ritz, K. (2005) Underview-origins and consequences of belowground biodiversity. In: *Biological Diversity and Function in Soil*, (eds R.D. Bardgett, M.B. Usher, and D.W. Hopkins), Cambridge University Press, Cambridge.

Ritz, K., McNicol, W., Nunan, N., Grayston, S., Millard, P., Atkinson, D., et al. (2004). Spatial structure in soil chemical and microbiological properties in an upland grassland. *FEMS Microbiology Ecology* 49(2), 191–205.

Roberts, M.S. and Cohen, F.M. (1995). Recombination and migration rates in natural populations of *Baccillus subtilus* and *Baccillus mojavensis*. *Evolution* 49, 1081–1094.

Robinson, C.H. and Wookey, P.A. (1997). Microbial ecology, decomposition and nutrient cycling. In: *Ecology and Arctic Environments* (eds. S.J. Woodin and M. Marquiss), Special Publication Number 13 of the British Ecological Society, Blackwell Science, Oxford, pp. 41–68.

Robinson, C.H., Ineson, P., Piearce, T.G., and Rowland, A.P. (1992). Nitrogen mobilization by earthworms in limed peat soils under *Picea sitchensis*. *Journal of Applied Ecology* 29, 226–237.

Robinson, C.H., Ineson, P., Piearce, T.G., and Parrington, J. (1996). Effects of earthworms on cation and phosphate mobilisation in limed peat soils under *Picea sitchensis*. *Forest Ecology and Management* 86, 253–258.

Robinson, C.H., Miller, J.P., and Deacon, L.J. (2005). Biodiversity of saprotrophic fungi in relation to their function: do fungi obey the rules? In: *Biological Diversity and Function in Soil* (eds. R.D. Bardgett, M.B. Usher, and D.W. Hopkins), Cambridge University Press, Cambridge, UK.

Ronn, R., Gavito, M., Larsen, J., Jakobsen, I., Frederiksen, H., and Christensen, S. (2002). Response of free-living soil protozoa and micro-organisms to elevated atmospheric CO_2 and presence of mycorrhiza. *Soil Biology and Biochemistry* 34, 923–932.

Ross, D.J., Tate, K.R., Newton, P.C.D., Wilde, R.H., and Clark, H. (2000). Carbon and nitrogen pools and mineralization in a grassland gley soil under elevated carbon dioxide at a natural CO_2 spring. *Global Change Biology* 6, 779–790.

Ross, D.J., Newton, P.C.D., and Tate, K.R. (2004). Elevated CO_2 effects on herbage and soil carbon and nitrogen pools and mineralization in a species-rich, grazed pasture on a seasonally dry sand. *Plant and Soil* 260, 183–196.

Rouhier, H. and Read, D.J. (1999). Plant and fungal responses to elevated atmospheric CO_2 in mycorrhizal seedlings of *Betula pendula*. *Environmental and Experimental Botany* 42, 231–241.

Rouhier, H., Billes, G., Elkohen, A., Mousseau, M., and Bottner, P. (1994). Effect of elevated CO_2 on carbon and nitrogen distribution within a tree (*Castanea sativa* Mill) soil system. *Plant and Soil* 162, 281–292.

Ruess, L., Michelsen, A., Schmidt, I.K., and Jonasson, S. (1999). Simulated climate change affecting micro-organisms, nematode density and biodiversity in subarctic soils. *Plant and Soil* 212, 63–73.

Ruess, R.W., Hendrick, R.L., and Bryant, J.P. (1998). Regulation of fine root dynamics by mammalian browsers in early successional Alaskan taiga forests. *Ecology* 79, 2706–2720.

Saetre, P. and Bääth, E. (2000). Spatial variation and patterns of soil microbial community structure in a mixed spruce-birch stand. *Soil Biology and Biochemistry* 30, 909–917.

Salamon, J.A., Schaefer, M., Alphei, J., Schmid, B., and Scheu, S. (2004). Effects of plant diversity on Collembola in an experimental grassland ecosystem. *Oikos* 106, 51–60.

Salt, D.T., Fenwick, P., and Whittaker, J.B. (1996). Interspecific herbivore interactions in a high CO_2 environment: root and shoot aphids feeding on *Cardamine*. *Oikos* 77, 326–330.

Sankaran, M. and Augustine, D.J. (2004). Large herbivores suppress decomposer abundance in a semiarid grazing ecosystem. *Ecology* 85, 1052–1061.

Santos, P.F., Phillips, J., and Whitford, W.G. (1981). The role of mites and nematodes in early stages of buried litter decomposition in a desert. *Ecology* 62, 664–669.

Satchell, J.E. (1967). Lumbricidae. In: *Soil Biology* (eds. A. Burgess and F. Raw), Academic Press, London, pp. 259–322.

Schädler, M., Jung, G., Brandl, R., and Auge, M. (2004). Secondary succession is influenced by belowground insect herbivory on a productive site. *Oecologia* 138, 242–252.

Schadt, C.W., Martin, A.P., Lipson, D.A., and Schmidt, S.K. (2003). Seasonal dynamics of previously unknown fungal lineages in tundra soils. *Science* 301, 1359–1361.

Schenk, U., Manderscheid, R., Hugen, J., and Weigel, H.J. (1995). Effects of CO_2 enrichment and intraspecific competition on biomass partitioning, nitrogen-content and microbial biomass carbon in soil of perennial ryegrass and white clover. *Journal of Experimental Botany* 46, 987–993.

Scheu, S. (1987). The role of substrate-feeding earthworms (Lumbricidae) for bioturbation in a beechwood soil. *Oecologia* 72, 192–196.

Scheu, S. (2003). Effects of earthworms on plant growth: patterns and perspectives. *Pedobiologia* 47, 846–856.

Scheu, S. and Parkinson, D. (1994). Effects of invasion of an aspen forest (Canada) by *Dendrobaena octaedra* (Lumbricidae) on plant growth. *Ecology* 75, 2348–2361.

Scheu, S. and Schultz, E. (1996). Secondary succession, soil formation and development of a diverse community of oribatids and saprophagous soil macro-invertebrates. *Biological Conservation* 5, 235–250.

Scheu, S. and Schaefer, M. (1998). Bottom-up control of the soil macrofauna community in a beechwood on limestone: manipulation of food sources. *Ecology* 79, 1573–1585.

Scheu, S. and Falca, M. (2000). The soil food web of two beech forests (*Fagus sylvatica*) of contrasting humus type: stable isotope analysis of macro- and a mesofauna dominated community. *Oecologia* 123, 285–296.

Scheu, S., Theenhaus, A., and Jones, T.H. (1999). Links between the detritivore and the herbivore system: effects of earthworms and Collembola on plant growth and aphid development. *Oecologia* 119, 541–551.

Schimel, J.P. (1995). Ecosystem consequences of microbial diversity and community structure. In: *Arctic and Alpine Biodiversity: Patterns, Causes, and Ecosystem Consequences* (eds. F.S. Chapin and C. Korner), Springer-Verlag, Berlin, pp. 239–254.

Schimel, J.P. and Chapin, F.S. (1996). Tundra plant uptake of amino acid and NH_4^+ nitrogen in situ: plants compete well for amino acid N. *Ecology* 77, 2142–2147.

Schimel, J.P. and Clein, J.S. (1996). Microbial response to freeze–thaw cycles in tundra and taiga soils. *Soil Biology and Biochemistry* 28, 1061–1066.

Schimel, J.P., Cates, R.G., and Ruess, R.W. (1998). The role of balsam poplar secondary chemicals in controlling soil nutrient dynamics through succession in the Alaskan taiga. *Biogeochemistry* 42, 221–234.

Schimel, J.P., Bilbrough, C., and Welker, J.M. (2004). Increased snow depth affects microbial activity and nitrogen mineralization in two Arctic tundra communities. *Soil Biology and Biochemistry* 36, 217–227.

Schimel, J.P., Bennett, J., and Frierer, N. (2005). Microbial community composition and soil nitrogen cycling: is there really a connection? In: *Biological Diversity and Function in Soil* (eds. R.D. Bardgett, M.B. Usher, and D.W. Hopkins), Cambridge University Press, Cambridge, UK.

Schindler, D.W., Curtis, P.J., Bayley, S.E., Parker, B.R., Beaty, K.G., and Stainton, M.P. (1997). Climate-induced changes in the dissolved organic carbon budgets of boreal lakes. *Biogeochemistry* 36, 9–28.

Schipper, L.A., Degens, B.P., Sparling, G.P., and Duncan, L.C. (2001). Changes in microbial heterotrophic diversity along five plant successional sequences. *Soil Biology and Biochemistry* 33, 2093–2104.

Schlesinger, W.H. (1991). *Biogeochemistry: An Analysis of Global Change*. Academic Press, San Diego, CA.

Schlesinger, W.H. and Lichter, J. (2001). Limited carbon storage in soil and litter of experimental forest plots under increased atmospheric CO_2. *Nature* 411, 466–469.

Schmidt, S.K. and Lipson, D.A. (2004). Microbial growth under the snow: implications for nutrient allelochemical availability in temperate soils. *Plant and Soil* 259, 1–7.

Schmidt, I.K., Jonasson, S., and Michelsen, A. (1999a). Mineralization and microbial immobilization of N and P in arctic soils in relation to season, temperature and nutrient amendment. *Applied Soil Ecology* 11, 147–160.

Schmidt, O., Scrimgeor, C.M., and Curry, J.P. (1999b). Carbon and nitrogen stable isotope ratios in body tissue and mucus of feeding and fasting earthworms (*Lumbricus festivus*). *Oecologia* 118, 9–15.

Schmidt, S.K., Lipson, D.A., Ley, R.E., Fisk, M.C., and West, A.E. (2004). Impacts of chronic nitrogen additions vary seasonally and by microbial functional group in tundra soils. *Biogeochemistry* 69, 1–17.

Schneider, K., Migge, S., Norton, R.A., Scheu, S., Langel, R., Reineking, A., et al. (2004). Trophic niche differentiation in soil microarthropods (Oribatida, Acari): evidence from stable isotope ratios ($^{15}N/^{14}N$). *Soil Biology and Biochemistry* 36, 1769–1774.

Schnürer, J., Clarholm, M., and Rosswall, T. (1986). Fungi, bacteria and protozoa in soil from 4 arable cropping systems. *Biology and Fertility of Soils* 2, 119–126.

Schortemeyer, M., Hartwig, U.A., Hendrey, G.R., and Sadowsky, M.J. (1996). Microbial community changes in the rhizospheres of white clover and perennial ryegrass exposed to free air carbon dioxide enrichment (FACE). *Soil Biology and Biochemistry* 28, 1717–1724.

Schortemeyer, M., Dijkstra, P., Johnson, D.W., and Drake, B.G. (2000). Effects of elevated atmospheric CO_2 concentration on C and N pools and rhizosphere processes in a Florida scrub oak community. *Global Change Biology* 6, 383–391.

Seastedt, T.R., Ramundo, R.A., and Hayes, D.C. (1988). Maximisation of densities of soil animals by foliage herbivory: empirical evidence, graphical and conceptual models. *Oikos* 51, 243–248.

Setälä, H. (1995). Growth of birch and pine seedlings in relation to grazing by soil fauna on ectomycorrhizal fungi. *Ecology* 76, 1844–1851.

Setälä, H. and Huhta, V. (1990). Evaluation of the soil fauna impact on decomposition in a simulated coniferous forest soil. *Biology and Fertility of Soils* 10, 163–169.

Setälä, H., and Huhta, V. (1991). Soil fauna increase *Betula pendula* growth: Laboratory experiments with coniferous forest floor. *Ecology* 72, 665–671.

Setälä, H. and McLean, M.A. (2004). Decomposition rate of organic substrates in relation to the species diversity of soil saprophytic fungi. *Oecologia* 139, 98–107.

Setälä, H., Tyynismaa, M., Martikainen, E., and Huhta, V. (1991). Mineralization of C, N and P in relation to decomposer community structure in coniferous forest soil. *Pedobiologia* 35, 285–296.

Setälä, H., Kulmala, P., Mikola, J., and Markkola, A.M. (1999). Influence of ectomycorrhiza on the structure of detrital food webs in pine rhizosphere. *Oikos* 87, 113–122.

Setälä, H., Berg, M.P., and Jones, T.H. (2005). Trophic structure and functional redundancy in soil communities. In: *Biological Diversity and Function in Soil* (Eds. R.D. Bardgett, M.B. Usher, and D.W. Hopkins), Cambridge University Press, Cambridge, UK.

Sharpley, A.N. and Syers, J.K. (1976). Potential role of earthworm casts for the phosphorus enrichment of run-off waters. *Soil Biology and Biochemistry* 8, 341–346.

Sharpley, A.N., Syers, J.K., and Springett, J.A. (1979). Effect of surface casting earthworms on the transport of phosphorus and nitrogen in surface runoff from pasture. *Soil Biology and Biochemistry* 11, 459–462.

Shaver, G.R., Johnson, L.C., Cades, D.H., Murray, G., Laundre, J.A., Rastetter, E.B., et al. (1998). Biomass accumulation and CO_2 flux in wet sedge tundras: responses to nutrients, temperature, and light. *Ecological Monographs* 68, 75–97.

Shaver, G.R., Canadell, J., Chapin, F.S., Gurevitch, J., Harte, J., Henry, G., et al. (2000). Global warming and terrestrial ecosystems: a conceptual framework for analysis. *BioScience* 50, 871–882.

Siepel, H. (1996). Biodiversity of soil microarthropods: the filtering of species. *Biodiversity and Conservation* 5, 251–260.

Siepel, H. and van de Bund, C.F. (1988). The influence of management practices on the microarthropod community of grassland. *Pedobiologia* 31, 339–354.

Six, J., Ogle, A.M., Breidt, F.J., Conant, R.T., Mosiers, A.R., and Paustian, K. (2004). The potential to mitigate global warming with no-tillage management is only realized when practised in the long term. *Global Change Biology* 10, 155–160.

Smith, H.G. (1996). Diversity of Antarctic terrestrial protozoa. *Biodiversity and Conservation* 5, 1379–1394.

Smith, P. (2004). Soil as carbon sinks: the global context. *Soil Use and Management* 20, 212–218.

Smith, K.A. and Conen, F. (2004). Impacts of land management on fluxes and trace greenhouse gases. *Soil Use and Management* 20, 255–263.

Smith, S.E. and Read, D.J. (1997). *Mycorrhizal Symbiosis*: 2nd Edition. Academic Press, London.

Smith, V.R. and Steenkamp, M. (1990). Climate change and its ecological implications at subantarctic island. *Oecologia* 85, 14–24.

Smith, R.S., Shiel, R.S., Bardgett, R.D., Millward, D., Corkhill, P., Rolph, G., et al. (2003). Diversification management of meadow grassland: plant species diversity and functional traits associated with change in meadow vegetation and soil microbial communities. *Journal of Applied Ecology* 40, 51–64.

Söderström, B., Bååth, E., and Lundgren, B. (1983). Decrease in soil microbial activity and biomass owing to nitrogen amendments. *Canadian Journal of Microbiology* 29, 1500–1506.

Soussana, J.F., Loiseau, P., Vuichard, N., Ceschia, E., Balesdent, J., Chevallier, T., et al. (2004). Carbon cycling and sequestration opportunities in temperate grasslands. *Soil Use and Management* 20, 219–230.

Sparling, G.P. and Tinker, P.B. (1978). Mycorrhizal infection in Pennine grassland. I. Levels of infection in the field. *Journal of Applied Ecology* 15, 943–950.

Springett, J.A. (1992). Distribution of lumbricid earthworms in New Zealand. *Soil Biology and Biochemistry* 24, 1377–1381.

Staddon, P.L., Jakonsen, I., and Blum, H. (2004). Nitrogen input mediates the effects of free-air CO_2 enrichment on mycorrhizal fungal abundance. *Global Change Biology* 10, 1687–1688.

Stadler, B., and Michalzik, B. (1998). Linking aphid honeydew, throughfall, and forest floor solution chemistry of Norway spruce. *Ecology Letters* 1, 13–16.

Standen, V. (1984). Production and diversity of enchytraeids, earthworms and plants in fertilized hay meadow plots. *Journal of Applied Ecology* 21, 293–312.

Stanton, N.L. (1979). Patterns of species diversity in temperate and tropical litter mites. *Ecology* 62, 295–304.

Starfield, A.M. and Chapin, F.S. (1996). Model of transient changes in arctic and boreal vegetation in response to climate and land use change. *Ecological Applications* 6, 842–864.

Stark, S., and Grellmann, D. (2002). Soil microbial responses to herbivory in an arctic tundra heath at two levels of nutrient availability. *Ecology* 83, 2736–2744.

Stark, S., Wardle, D.A., Ohtonen, R., Helle, T., and Yeates, G.W. (2000). The effect of reindeer grazing on decomposition, mineralisation and soil biota in a dry oligotrophic Scots pine forest. *Oikos* 90, 301–310.

Stark, S., Tuomi, J., Strömmer, R., and Helle, T. (2003). Non-parallel changes in soil microbial carbon and nitrogen dynamics due to reindeer grazing in northern boreal forests. *Ecography* 26, 51–59.

Ştefan, V. (1977). Soil Enchytraeidae from the Cerna Valley. Fourth Symposium on Soil Biology; Rumanian National Society of Soil Science. Editura Ceres, Bucureşti, pp. 277–283.

Steltzer, H. and Bowman, W.D. (1998). Differential influence of plant species on soil N transformations within moist meadow alpine tundra. *Ecosystems* 1, 464–474

Stevens, C.J., Dise, N.B., Mountford, J.O., and Gowing, D.J. (2004). Impact of nitrogen deposition on the species richness of grasslands. *Science* 303, 1876–1879.

Stocker, R., Körner, C., Schmid, B., Niklaus, P.A., and Leadley, P.W. (1999). A field study of the effects of elevated CO_2 and plant species diversity on ecosystem-level gas exchange in a planted calcareous grassland. *Global Change Biology* 5, 95–105.

Streeter, T.C., Bol, R., and Bardgett, R.D. (2000). Amino acids as a nitrogen source in temperate upland grasslands: the use of dual labelled (^{13}C, ^{15}N) glycine to test for direct uptake by dominant grasses. *Rapid Communications in Mass Spectrometry* 14, 1351–1355.

Subler, S., Baranski, C.M., and Edwards, C.A. (1997). Earthworm additions increased short-term nitrogen availability and leaching in two grain-crop agroecosystems. *Soil Biology and Biochemistry* 29, 413–421.

Sugden, A., Stone, R., and Ash, C. (2004). Ecology in the underworld. *Science* 304, 1613.

Swift, M.J., Heal, O.W., and Anderson, J.M. (1979). *Decomposition in Terrestrial Ecosystems*. University of California Press, Berkeley.

Teuben, A. (1991). Nutrient availability and interactions between soil arthropods and microorganisms during decomposition of coniferous litter: a mesocosm study. *Biology and Fertility of Soils* 10, 256–266.

Thompson, C.H. (1981). Podzol chronosequences on coastal dunes of eastern Australia. *Nature* 291, 59–61.

Thompson, C.H. (1992). Genesis of podzols on coastal dunes in southern Queensland. I. Field relationships and profile morphology. *Australian Journal of Soil Research* 30, 593–613.

Tian, G., Brussaard, L., and Kang, B.T. (1992). Biological effects of plant residues with contrasting chemical composition under humid tropical conditions: decomposition and nutrient release. *Soil Biology and Biochemistry* 24, 1051–1060.

Tilman, D. (1982). *Resource Competition and Community Structure*. Princeton University Press, Princeton, USA.

Tinker, P.B. and Ineson, P. (1990). Soil organic matter and biology in relation to climate change. *Developments in Soil Science* 20, 71–88.

Tipping, E., Woof, C., Rigg, E., Harrison, A.F., Ineson, P., Taylor, K., et al. (1999). Climatic influences on the leaching of dissolved organic matter from upland UK moorland soils, investigated by a field manipulation experiment. *Environment International* 25, 83–95.

Torsvik, V., Ovreas, L., and Thingstad, T.F. (2002). Prokaryotic diversity: magnitude, dynamics, and controlling factors. *Science* 296, 1064–1066.

Tracey, B.F. and Frank, D.A. (1998). Herbivore influence on soil microbial biomass and N mineralization in a northern grassland ecosystem: Yellowstone National Park. *Oecologia* 114, 556–562.

Trumbore, S.E., Chadwick, O.A., and Amundson, R. (1996). Rapid exchange between soil carbon and atmospheric carbon dioxide driven by climate change. *Science* 272, 393–396.

Tunlid, A. and White, D.C. (1992). Biochemical analysis of biomass, community structure, nutritional status, and metabolic activity of microbial communities in soil. In: *Soil Biochemistry* (eds. G. Stotzky and J.M. Bollag), Vol 7, Marcel Dekker, NY, pp. 229–262.

Turner, B.L. and Haygarth, P.M. (2001). Phosphorus solubilisation in rewetted soils. *Nature* 411, 258.

Turner, B.L., Papházy, M.J., Haygarth, P.M., and McKelvie, I.D. (2002). Inositol phosphates in the environment. *Philosophical Transactions of the Royal Scoeity, London* 357, 449–469.

Turner, B.L., Frossard, E., and Baldwin, D.S. (eds.) (2004*a*). *Organic Phosphorus in the Environment*. CABI International, Wallingford.

Turner, B.L., Baxter, R., Mahieu, N., Sjőgersten, S., and Whitton, B. (2004*b*). Phosphorus compounds in subarctic Fennoscandian soils at the mountain birch (*Betula pubesecens*)-tundra ecotone. *Soil Biology and Biochemistry* 36, 815–823.

Van Dam, D., Van Dobben, H.F., Ter Braak, C.J.F., and DeWit, T. (1986). Air pollution as a possible cause for the decline of some phanerogamic species in the Netherlands. *Vegetatio* 65, 47–52.

Van Dam, N.M., Horn, M., Mares, M., and Baldwin, I.T. (2001). Ontogeny constrains systematic protease inhibitor response in *Nictotiana attenuate*. *Journal of Chemical Ecology* 27, 547–568.

Van Dam, N.H., Harvey, J.A., Wäckers, F.L., Bezemer, T.M., Van der Putten, W.H., and Vet, L.E.M. (2003). Interactions between aboveground and belowground induced responses against phytophages. *Basic and Applied Ecology* 4, 63–77.

Van der Heijden, M.G.A. (2004). Arbuscular mycorrhizal fungi as support systems for seedling establishment in grassland. *Ecology Letters* 7, 293–303.

Van der Heijden, M.G.A., Boller, T., Wiemken, A., and Sanders, I.R. (1998a). Different arbuscular mycorrhizal fungal species are potential determinants of plant community structure. *Ecology* 79, 2082–2091.

Van der Heijden, M.G.A., Klironomos, J.N., Ursic, M., Moutoglis, P., Streitwolf-Engel, R., Boller, R., et al. (1998b). Mycorrhizal fungal diversity determines plant biodiversity, ecosystem variability and productivity. *Nature* 396, 69–72.

Van der Putten, W.H. (2005). Plant-soil feedback and soil biodiversity affect the composition of plant communities. In: *Biological Diversity and Function in Soil* (eds. R.D. Bardgett, M.B. Usher, and D.W. Hopkins), Cambridge University Press, Cambridge, UK.

Van der Putten, W.H., Van Dijk, C., and Peters, B.A.M. (1993). Plant-specific soil-borne diseases contribute to succession in foredune vegetation. *Nature* 362, 53–56.

Van der Wal, R. and Brooker, R.W. (2004). Mosses mediate grazer impacts on grass abundance in arctic ecosystems. *Functional Ecology* 18, 77–86.

Van der Wal, R., Van Wijnen, H., Van Wieren, S., Beucher, O., and Bos, D. (2000). On facilitation between herbivores: how Brent geese profit from brown hares. *Ecology* 81, 969–980.

Van der Wal, R., Pearce, I.S.K., Brooker, R., Scott, D., Welch, D., and Woodin, S.J. (2003). Interplay between nitrogen deposition and grazing causes habitat degradation. *Ecology Letters* 6, 141–146.

Van der Wal, R., Bardgett, R.D., Harrison, K.A., and Stien, A. (2004). Vertebrate herbivores and ecosystem control: cascading effects of faeces on tundra ecosystems. *Ecography* 27, 242–252.

Van der Wal, R., Pearce, I.S.K., and Brooker, R.W. (2005). Mosses and the struggle for light in a nitrogen polluted world. *Oecologia,* 142, 159–168.

Vandenkoornhuyse, P., Husband, R., Daniell, T.J., Watson, I.J., Duck, J.M., Fitter, A.H., et al. (2002). Arbuscular mycorrhizal community composition associated with two plant species in a grassland ecosystem. *Molecular Ecology* 11, 1555–1564.

Väre, H., Ohtonen, R., and Mikkola, K. (1996). The effect and extent of heavy grazing by reindeer in oligotrophic pine heaths in northeastern Fennoscandia. *Ecography* 19, 245–253.

Vazquez, D.P. (2002). Multiple effects of introduced mammalian herbivores in a temperate forest. *Biological Invasions* 4, 175–191.

Verhoef, H.A. and van Selm, A.J. (1983). Distribution and population dynamics of Collembola in relation to soil moisture. *Holarctic Ecology* 6, 387–394.

Verity, P.G. (1991). Feeding in planktonic protozoans—evidence for non-random acquisition of prey. *Journal of Protozoology* 38, 69–76.

Verrecchia, E., Yair, A., Kidron, G.J., and Verrecchia, K. (1995). Physical properties of the psammophile crpytogamic crusts and their consequences to the water regime of sandy soils, northwestern Negev, Israel. *Journal of Arid Environments* 29, 427–437.

Verschoor, B.C. (2001). *Nematode-Plant Interactions in Grasslands under Restoration Management.* PhD Thesis, University of Wageningen, The Netherlands.

Virginia, R.A. and Wall, D.H. (1999). How soils structure communities in the Antarctic Dry Valleys. *BioScience* 49, 973–983.

Virtanen, R. (2000). Effects of grazing on above-ground biomass on a mountain snowbed, NW Finland. *Oikos* 90, 295–300.

Visser, S., Whittaker, J.B., and Parkinson, D. (1981). Effects of collembolan grazing on nutrient release and respiration of a leaf litter inhabiting fungus. *Soil Biology and Biochemistry* 13, 215–218.

Vitousek, P.M. (1990). Biological invasions and ecosystem processes: towards an integration of population biology and ecosystem studies. *Oikos* 57, 7013.

Vitousek, P.M. and Hobbie, A. (2000). Heterotrophic nitrogen-fixation in decomposing litter: patterns and regulation. *Ecology* 81, 2366–2376.

Vitousek, P.M. and Walker, L.R. (1989). Biological invasion by *Myrica faya* in Hawai'i: plant demography, nitrogen fixation, ecosystem effects. *Ecological Monographs* 59, 247–265.

Vitousek, P.M., Walker, L.R., Whittaker, L., Mueller-Dombois, D., and Matson, P. (1987). Biological invasion of *Myrica faya* alters ecosystem development in Hawaii. *Science* 238, 802–804.

Vitousek, P.M., Mooney, H.A., Lubchenco, J., and Melilllo, J.M. (1997). Human domination of Earth's ecosystems. *Science* 277, 494–499.

Vitousek, P.M., Cassman, K., Cleveland, C., Crews, T., Field, C.B., Grimm, N.B., et al. (2002). Towards an ecological understanding of biological nitrogen fixation. *Biogeochemistry* 57, 1–45.

Volz, P. (1951). Unterschungen uber die Microfauna des Waldbodens. Zool. Jahrb. Abt. Syst. Oekol. Geogr. *Tiere* 79, 514–566.

Waldrop, M.P., Zak, D.R., and Sinsabaugh, R.L. (2004). Microbial community responses to nitrogen deposition in northern forest ecosystems. *Soil Biology and Biochemistry* 36, 1443–1451.

Walker, L.R. (1993). Nitrogen fixers and species replacements in primary succession. In: *Primary Succession on Land* (eds. J. Miles and D.W.H. Walton), Blackwell Scientific, Oxford, pp. 249–272.

Walker, L.R. and Del Moral, R. (2003). *Primary Succession and Ecosystem Rehabilitation.* Cambridge University Press, Cambridge, UK.

Walker, T. and Syers, J.K. (1976). The fate of phosphorus during pedogensis. *Geoderma* 15, 1–19.

Walker, M.D., Walker, D.A., Welker, J.M., Arft, A.M., Bardsley, T., Brooks, P.D., et al. (1999). Long-term experimental manipulation of winter snow regime and summer temperature in arctic and alpine tundra. *Hydrological Processes* 13, 2315–2330.

Wall, D.H. and Virginia, R.A. (1999). Controls on soil biodiversity: insights from extreme environments. *Applied Soil Ecology* 13, 137–150.

Wall, J.W., Skene, K.R., and Neilsen, R. (2002). Nematode community and trophic structure along a sand dune succession. *Biology and Fertility of Soils* 35, 293–301.

Walsingham, J.M. (1976). Effect of sheep grazing on the invertebrate population of agricultural grassland. *Proceedings of the Royal Society of Dublin* 11, 297–304.

Wardle, D.A. (1992). A comparative assessment of factors which influence microbial biomass carbon and nitrogen levels in soils. *Biological Reviews* 67, 321–358.

Wardle, D.A. (2002). *Communities and Ecosystems: Linking the Aboveground and Belowground Components.* Monographs in Population Biology 34. Princeton University Press, NJ.

Wardle, D.A. (2005). How plant communities influence decomposer communities. In: *Biological Diversity and Function in Soil,* (eds. R.D. Bardgett, M.B. Usher, and D.W. Hopkins), Cambridge University Press, Cambridge, UK.

Wardle, D.A. and Nicholson, K.S.(1996). Synergistic effects of grassland plant species on soil microbial biomass and activity: implications for ecosystem-level effects of enriched plant diversity. *Functional Ecology* 10, 410–416.

Wardle, D.A. and Van der Putten, W.H. (2002). Biodiversity, ecosystem functioning and above-ground-below-ground linkages. In: *Biodiversity and Ecosystem Functining* (eds. M. Loreaui, S. Naeem, and P. Inchausti), Oxford University Press, Oxford, pp 155–168.

Wardle, D.A. and Bardgett, R.D. (2004a). Human-induced changes in densities of large herbivorous mammals: consequences for the decomposer subsystem. *Frontiers in Ecology and the Environment* 2, 145–153.

Wardle, D.A. and Bardgett, R.D. (2004b). Indirect effects of invertebrate herbivory on the decomposer subsystem. In: *Insects and Ecosystem Functioning* (eds. W. Weisser and E. Siemann). Ecological Studies 173. Springer, Heidelberg, pp. 53–69.

Wardle, D.A., Zackrisson, O., Hörnberg, G., and Gallet, C. (1997a). The influence of island area on ecosystem properties. *Science* 277, 1296–1299.

Wardle, D.A., Bonner, K.I., and Nicholson, K.S. (1997b). Biodiversity and plant litter: experimental evidence which does not support the view that enhanced species richness improves ecosystem function. *Oikos* 79, 247–258.

Wardle, D.A., Barker, G.M., Bonner, K.I., and Nicholson, K.S. (1998). Can comparative approaches based on plant ecophysiological traits predict the nature of biotic interactions and individual plant species effects in ecosystems? *Journal of Ecology* 86, 405–420.

Wardle, D.A., Bonner, K.L., Barker, G.M., Yeates, G.W., Nicholson, K.S., Bardgett, R.D., et al. (1999a). Plant removals in perennial grassland: vegetation dynamics, decomposers, soil biodiversity, and ecosystem properties. *Ecological Monographs* 69, 535–568.

Wardle, D.A., Nicholson, K.S., Bonner, K.I., and Yeates, G.W. (1999b). Effects of agricultural intensification on soil-associated arthropod population dynamics, community structure, diversity and temporal variability over a seven-year period. *Soil Biology and Biochemistry* 31, 1691–1706.

Wardle, D.A., Barker, G.M., Yeates, G.W., Bonner, K.I., and Ghani, A. (2001). Impacts of introduced browsing mammals in New Zealand forests on decomposer communities, soil biodiversity and ecosystem properties. *Ecological Monographs* 71, 587–614.

Wardle, D.A., Yeates, G.W., Barker, G.M., Bellingham, P.J., Bonner, K.I., and Williamson, W.M. (2003a). Island biology and ecosystem functioning in epiphytic soil communities. *Science* 301, 1717–1720.

Wardle, D.A., Yeates, G.W., Williamson, W., and Bonner, K. (2003b). The response of three trophic level soil food web to the identity and diversity of plant species and functional groups. *Oikos* 102, 45–56.

Wardle, D.A., Bardgett, R.D., Klironomos, J.N., Setälä, H., van der Putten, W.H., and Wall, D.H. (2004*a*). Ecological linkages between aboveground and belowground biota, *Science* 304, 1629–1633.

Wardle, D.A., Walker, L.R., and Bardgett, R.D. (2004*b*). Ecosystem properties and forest decline in contrasting long-term chronosequences. *Science* 305, 509–513.

Warnock, A.J., Fitter, A.H., and Usher, M.B. (1982). The influence of a springtail *Folsomia candida* (Insecta, Collembola) on the mycorrhizal association of leek *Allium porrum* and the vesicular-arbuscular mycorrhizal endophyte *Glomus fasciculatum*. *New Phytologist* 90, 285–292.

Watson, R. (1999). Common themes for ecologists in global issues. *Journal of Applied Ecology* 36, 1–10.

Webb, D.P. (1977). Regulation of deciduous forest litter decomposition by soil arthropod feces. In: *The Role of Arthropods in Forest Ecosystems* (ed. W.J. Mattson), Springer-Verlag. pp. 57–69.

Webb, N.R., Coulson, S.J., Hodkinson, I.D., Block, W., Bale, J.S., and Strathdee, A.T. (1998). The effects of experimental temperature elevation on populations of cryptostigmatic mites in high Arctic soils. *Pedobiologia* 42, 298–308.

Wedin, D.A. and Tilman, D. (1996). Influence of nitrogen loading and species composition on the carbon balance of grasslands. *Science* 274, 1720–1723.

Weigelt, A., King, R., Bol, R., and Bardgett, R.D. (2003). Inter-specific variability in the uptake of organic nitrogen in three dominant grasses of temperate grassland. *Soil Science and Plant Nutrition* 166, 1–6.

Weigelt, A., Bol, R., and Bardgett, R.D. (2005). Preferential uptake of soil nitrogen forms by grassland plant species. *Oecologia,* 142, 627–635.

Weis-Fogh, T. (1948). Ecological investigations of mites and collemboles in the soil. App. Description of some new mites (Acari). *Natura Jutlandica* 1, 139–277.

Welker, J.M., Bowman, W.D., and Seastedt, T.R. (2001). Environmental change and future directions in alpine research. In: *Structure and Function of an Alpine Ecosystem* (eds. W.D. Bowman and T.R. Seastedt), Oxford University Press, Oxford, pp. 304–322.

Welker, J.M., Brown, K., and Fahnestock, J.T. (1999). CO_2 flux in arctic and alpine dry tundra: comparative field responses under ambient and experimentally warmed conditions. *Arctic, Antarctic, and Alpine Research* 31, 272–277.

Welker, J.M., Molau, U., Parsons, A.N., Robinson, C.H., and Wookey, P.A. (1997). Responses of *Dryas octopetela* to ITEX environmental manipulations: a synthesis with circumpolar comparisons. *Global Change Biology* 3, 61–73.

West, T.O. and Marland, G. (2002). A synthesis of carbon sequestration, carbon emissions, and net carbon flux in agriculture: comparing tillage practices in the United States. *Agriculture, Ecosystems and Environment* 91, 217–232.

Westover, K.M. and Bever, J.D. (2001). Mechanisms of plant species coexistence: roles of rhizosphere bacteria and root fungal pathogens. *Ecology* 82, 3285–3294.

Whipps, J.M. and Lynch, J.M. (1983). Substrate flow and utilization in the rhizosphere of cereals. *New Phytologist* 95, 605–623.

White, D.C., Davis, W.M., Nickels, J.S., King, J.C., and Bobbie, R.J. (1979). Determination of the sedimentary microbial biomass by extractable lipid phosphate. *Oecologia* 40, 51–62.

White, R.E. (1997). *Principles and Practice of Soil Science*, 3rd edn., Blackwell Science, Oxford.

Wilson, E.O. (2002). *The Future of Life*. Alfred A. Knof, New York.

Woese, C.R. (1987). Bacterial evolution. *Microbiology Review* 51, 221–271.

Woods, L.E., Cole, C.V., Elliott, E.T., Anderson, R.V., and Coleman, D.C. (1982). Nitrogen transformations in soil as affected by bacterial–microfaunal interactions. *Soil Biology and Biochemistry* 14, 93–98.

Woodell, S.J. and King, T.J. (1991). The influence of mound-building ants on British lowland vegetation. In: *Ant-Plant Interactions* (eds. C.R. Huxley, and D.F. Cutler), Oxford University Press, Oxford, pp. 521–535.

Woodin, S.J. and Lee, J.A. (1987). The effects of nitrate, ammonium and temperature on nitrate reductase activity in *Sphagnum* species. *New Phytologist* 105, 103–105.

Worral, F., Burt, T., and Shedden, R. (2003). Long term records of riverine dissolved organic matter. *Biogeochemistry* 64, 165–178.

Worral, F., Burt, T., and Adamson, J. (2004). Can climate change explain increases in DOC flux from upland peat catchments? *Science of the Total Environment* 326, 95–112.

Wurst, S., Langel, R., Reineking, A., Bonkowski, M., and Shceu, S. (2003). Effects of earthworms and organic litter distribution on plant performance and aphid reproduction. *Oecologia* 137, 90–96.

Wynn-Williams, D.D. (1993). Microbial processes and intial stabilization of fellfield soil. In: *Primary Succession on Land* (eds. J. Miles and D.W.H. Walton), Blackwell Scientific, Oxford, pp. 17–32.

Yeates, G.W. (1968). An analysis of annual variation of the nematode fauna in dune sand, at Himatangi Beach, New Zealand. *Pedobiologia* 8, 173–207.

Yeates, G.W. and King, K.L. (1997). Soil nematodes as indicators of the effect of management on grasslands in the New England Tablelands (NSW): comparison of native and improved grasslands. *Pedobiologia* 41, 526–536.

Yeates, G.W., Bongers, T., De Goede, R.G.M., Freckman, D.W., and Georgieva, S.S. (1993). Feeding habits in soil nematode families and genera—an outline for ecologists. *Journal of Nematology* 25, 315–331.

Yeates, G.W., Bardgett, R.D., Cook, R., Hobbs, P.J., Bowling, P.J, and Potter, J.F. (1997a). Faunal and microbial diversity in three Welsh grassland soils under conventional and organic management regimes. *Journal of Applied Ecology* 34, 453–471.

Yeates, G.W., Tate, K.R., and Newton, P.C.D. (1997b). Response of the fauna of a grassland soil to doubling of atmospheric carbon dioxide concentration. *Biology and Fertility of Soils* 25, 307–315.

Yeates, G.W., Saggar, S., Denton, C.S. and Mercer, C.F. (1998). Impact of clover cyst nematode (*Heterodera trifolii*) infection on soil microbial activity in the rhizosphere of white clover (*Trifolium repens*)—a pulse labelling experiment. *Nematologica* 44, 81–90.

Yeates, G.W., Hawke, M.F., and Rijske, W.C. (2000). Changes in soil fauna and soil conditions under *Pinus radiata* agroforestry regimes during a 25-year tree rotation. *Biology and Fertility of Soils* 30, 391–406.

Young, I.M. and Crawford, J.W. (2004). Interactions and self-organization in the soil-microbe complex. *Science* 304, 1634–1637.

Zak, D.R., Pregitzer, K.S., Curtis, P.S., Teeri, J. A., Fogel, R., and Randlett, D.L. (1993). Elevated atmospheric CO_2 and feedback between carbon and nitrogen cycles. *Plant and Soil* 151, 105–117.

Zak, D.R., Ringelberg, D.B., Pregitzer, K.S., Randlett, D.L., White, D.C., and Curtis, P.S. (1996). Soil microbial communities beneath *Populus grandidentata* crown under elevated atmospheric CO_2. *Ecological Applications* 6, 257–262.

Zak, D.R., Pregitzer, K.S., Curtis, P.S., Vogel, C.S., Holmes, W.E., and Lussenhop, J. (2000). Atmospheric CO_2, soil-N availability, and allocation of biomass and nitrogen by *Populus tremuloides*. *Ecological Applications* 10, 34–46.

Zak, D.R., Holmes, W.E., Finzi, A.C., Norby, R.J., and Schlesinger, W.H. (2003*a*). Soil nitrogen cycling under elevated CO_2: a synthesis of forest FACE experiments. *Ecological Application* 13, 1508–1514.

Zak, D.R., Holmes, W.E., White, D.C., Peacock, A.D., and Tilman, D. (2003*b*). Plant diversity, soil microbial communities, and ecosystem function: are there any links? *Ecology* 84, 2042–2050.

Zhang, H. and Forde, B.G. (1998). An Arabidopsis MADS box gene that controls nutrient-induced changes in root architecture. *Science* 279, 407–409.

Zhou, J., Xia, B., Treves, D.S., Wu, L.Y., Marsh, T.L., O'Neill, R.V., et al. (2002). Spatial and resource factors influencing high microbial diversity in soil. *Applied Environmental Microbiology* 68, 326–334.

Zhu, Y. G., Laidlaw, A.S., Christie, P., and Hammond, M.E.R. (2000). The specificity of arbuscular mycorrhizal fungi in perennial ryegrass-white clover pasture. *Agriculture, Ecosystems and Environment* 77, 211–218.

Zimmer, M. and Topp, W. (1998). Microorganisms and cellulose digestion in the gut of *Porcellio scaber* (Isopoda: Oniscidea). *Journal of Chemical Ecology* 24, 1397–1408.

Zimmer, M. and Topp, W. (2002). The role of coprophagy in nutrient release from feces of phytophagous insects. *Soil Biology and Biochemistry* 34, 1093–1099.

Zimov, S.A., Chuprynin, V.I., Oreshko, A.P., Chapin, F.S. III, Reynolds, J.F., and Chapin, M.C. (1995). Steppe-tundra transition: a herbivore-driven biome shift at the end of the Pleistocent. *American Naturalist* 146, 765–794.

Zogg, D.G., Zak, D.R., Pregitzer, K.S., and Burton, A.J. (2000). Microbial immobilization and the retention of anthropogenic nitrate in a northern hardwood forest. *Ecology* 81, 1858–1866.

Index

Note: page numbers in *italics* refer to boxes, figures and tables.